Praise for *Meat*

"Bruce's book is an impressive *tour d'horizon* of the need for alternatives to animal-based protein and how we can make alt-meats—with taste and price parity—a reality . . . This book explains the imperative to transform our food systems and lays out a game plan to get us there . . . *Meat* is as important as it is enjoyable. I hope you'll be as inspired to help create an alt-meat future as I am."

—Caitlin Welsh, director of the Global Food and Water Security Program at the Center for Strategic and International Studies

"Bruce Friedrich's *Meat* is an engaging treatise on using science to make meat far more efficiently. If it works, and Friedrich is convincing that science is up to the challenge, these alternative meats will slash humanity's adverse impact on our natural world, including our oceans, forests, and climate. Friedrich includes fascinating observations in every chapter, including his analysis of 'ultra-processed foods' that are actually super-healthy and a final chapter that is a compelling and practical guide to what each of us can do to help."

—George Church, professor of genetics at Harvard Medical School, lead for synthetic biology at Wyss Institute of Biologically Inspired Engineering, and author of *Genesis*

"Terrific, fun, humane, and inspiring. Friedrich shows that a better future, with alternative meats, is coming—and that we're going to love what we eat."

—Cass R. Sunstein, Robert Walmsley University professor at Harvard University and author *of How Change Happens*

"Today billions of animals are raised to provide us with meat. This harms the environment and is often unspeakably cruel to farmed animals, yet for many the idea of a main meal without meat is unthinkable. The good news, as *Meat* explains, is that you can now enjoy real meat that was made without killing any animal. Please read this book: it is engaging, informative, and gives us hope for a kinder future."

—Jane Goodall, PhD, DBE, founder of the Jane Goodall Institute and UN Messenger of Peace

"This book is an eye-opener that could help save the world . . . The topic is crucial, and Friedrich's presentation is clear, persuasive, and entertaining. Meat of all kinds will never look or taste the same—along with everything else, you can eat with a dash of hope to taste."

—Kim Stanley Robinson, winner of the Hugo, Nebula, and Locus Awards
and author of *The Ministry for the Future*

"Alternative proteins offer a promising path to addressing global challenges including hunger, climate change, and pandemic risk. *Meat* highlights the case for how science, innovation, and smart policy can help bring these solutions within reach. It contributes to an important and timely global conversation."

—Michael Kremer, professor at the University of Chicago
and corecipient of the 2019 Nobel Memorial Prize in Economic Sciences

"I really love that *Meat* invites everyone to the table. The alt-meat scientific challenge has support from a broad array of Republicans and Democrats in the US, as well as from policymakers globally across the range of ideologies. What's more, the current meat industry will be, according to Bruce Friedrich, a big part of the solution. Equal parts urgent and optimistic, *Meat* is meticulously researched, fast-paced and fun to read, and full of the kind of ideas that could actually work. A landmark contribution to the future of food."

—Paul Wesley, actor (*Star Trek: Strange New Worlds; The Vampire Diaries*)

"Our food system is both an overlooked national security vulnerability and a powerful catalyst for the next wave of American innovation. In *Meat*, Bruce Friedrich reveals the fragile underpinnings of global meat production and the immense risks it poses to geopolitical stability, public health, and the environment. With rigorous research, sharp economic analysis, and real-world stories that span boardrooms and battlefields, Friedrich rightly reframes the problem as a generational opportunity. This is the rare book that speaks to investors, policymakers, technologists—and anyone who eats."

—Matt Spence, global head of venture capital and managing director at Barclays,
deputy assistant secretary of defense for Middle East Policy (2012–15),
and member of the US National Security Council (2009-2012)

"In *Meat*, Bruce Friedrich makes a compelling case for alternative proteins as a globally scalable solution to some of our most urgent public global health and ecological crises. He also shows how we get there: by focusing on the economic and food security benefits of protein transition. The best global health policy book of the decade turns out to be about remaking meat. Essential reading for policymakers, and for anyone who eats."

—Michael Greger, MD, founder of NutritionFacts.org and author of *How Not to Die*
and *Bird Flu: A Virus of Our Own Hatching*

"*Meat* is a fascinating exploration of the consequences of modern meat production and a road map to a better future for meat manufacturing. With clarity and conviction, *Meat* charts a path toward a world where our appetites and our ethics can finally align."

—Steve Jurvetson, founder and managing director of Future Ventures
and board member of SpaceX

"*Meat* offers a bold reimagining of how we can make the 'meat' people love—more affordably, more sustainably, and more equitably. With insight and wit, Bruce Friedrich explores diverse solutions that preserve cultural traditions and taste while confronting the urgent challenges of climate change, global health, and food insecurity. "
—Ertharin Cousin, founder and CEO of Food Systems for the Future,
visiting scholar at the Stanford University Center on Food Security and Environment,
and 2012–2017 executive director of the United Nations World Food Programme

"In this important book, Bruce Friedrich shares his expertise and his passion for making meat better. He makes a powerful case that, with innovation and investment, we can meet the world's growing demand for meat that is delicious, nutritious, and affordable while also protecting human and planetary health. A fascinating and inspiring read!"
—Charlotte Pera, executive director of Stanford University's Sustainability Accelerator
and senior fellow of the Bezos Earth Fund

"An unflinching case for reinventing meat, so that we can feed a growing world without feeding the climate crisis."
—Christiana Figueres, chair of the Earthshot Prize Foundation,
cofounder of Global Optimism, and 2010–2016 executive secretary
of the UN Framework Convention on Climate Change

"Friedrich tells an engaging story of the visionaries who are inventing a new way to produce meat by combining proteins, fats, and nutrients. The idea might sound far-fetched today, but as he shows in a quick tour of past breakthroughs—cars, airplanes, computers, smartphones—what once seemed unimaginable can soon become commonplace."
—Kathryn Aschheim, deputy editor of *Nature Biotechnology*

"Bruce Friedrich has a unique talent for making radical change feel not only possible, but inspiring and near at hand. Our future depends on people like him, and books like this."
—Jonathan Safran Foer, Lillian Vernon Distinguished writer-in-residence
at New York University and author of *Everything Is Illuminated*
and *Extremely Loud and Incredibly Close*

"In 2018, I accidentally heard a talk by Bruce Friedrich that fundamentally changed my eating, teaching, philanthropy, and investing. Bruce provided compelling logic on how we can improve the environment, our health, and animal welfare, as well as how to feed the world's population. This book provides broader access to his vision on how to make the world better. But be careful; reading this book could change your life."
—Max Bazerman, Straus Professor at Harvard Business School
and author of *Inside an Academic Scandal* and *Negotiation: The Game Has Changed*

"Bruce Friedrich is a close friend of mine. Years ago he took me out to dinner to persuade me that plant-based meat is just as succulent as beef, and is the inevitable future of haute cuisine. We did a blind taste test. His contention fell apart, miserably, and I joyfully wrote about it in *The Washington Post*. I am now convinced he wrote this book entirely to prove to me that I was a blind, ignorant, cynical, arrogant fool who understood nothing about alternative meats and their potential. So I just read it. He is right."

—Gene Weingarten, two-time Pulitzer Prize winner
for feature writing at *The Washington Post* and author of
One Day: The Extraordinary Story of an Ordinary 24 Hours in America

"The heart of this book is a fascinating story of scientific and entrepreneurial exploration. Trying to alter ancient eating habits is hard, but perhaps not impossible. And it would be extremely useful for the world if they can bring this tale to a successful climax."

—Bill McKibben, author of *Here Comes the Sun*

"Bruce Friedrich shows why, if we are to avoid pandemics, feed the poor, and mitigate the severity of climate change, we need to scale up the production of meat that does not come from animals. He also gives us hope that this is possible."

—Peter Singer, professor emeritus of bioethics at Princeton University
and author of *Animal Liberation*

Praise for the Good Food Institute

"GFI continues to be a powerhouse in alternative protein thought leadership and action. It has strong ties to government, industry, and research organizations and continues to achieve impressive wins. We believe donations to GFI can help stimulate systemic change that reduces food system emissions on a global scale."

—Giving Green, charity evaluator

"The Good Food Institute has been the global catalyst for alternative protein innovation, centering alternative proteins as an essential food and land use climate solution . . . The Bezos Earth Fund supports GFI because we recognize their vital role as a field catalyst for alternative proteins. We encourage others to join us in this support to fuel even greater impact."

—Andy Jarvis, director of the future of food at the Bezos Earth Fund

"The Good Food Institute is . . . second-to-none in the influence of its public policy efforts, its centrality to the ecosystem of companies and researchers, and its international footprint. It has also been effective at convincing traditional meat companies to explore alternative proteins, which could lead both to important products and turn political enemies into allies."

—Ezra Klein, *New York Times* columnist

MEAT

How the Next
Agricultural Revolution
Will Transform
Humanity's Favorite Food—
and Our Future

BRUCE FRIEDRICH

BENBELLA

BenBella Books, Inc.
Dallas, TX

to Alka, for everything

BenBella

BenBella Books, Inc.
8080 N. Central Expressway
Suite 1700
Dallas, TX 75206
benbellabooks.com
Send feedback to feedback@benbellabooks.com

BenBella is a federally registered trademark.

Printed in the United States of America
10 9 8 7 6 5 4 3 2

Library of Congress Control Number: 2025033537
ISBN 9781637747933 (hardcover)
ISBN 9781637747940 (electronic)

Editing by Greg Brown
Copyediting by Kaya Skovdatter
Proofreading by Martha Gallant and Cheryl Beacham
Indexing by WordCo Indexing Services
Text design and composition by Jordan Koluch
Cover design by Brigid Pearson
Cover image © Adobe Stock / Andrey
Printed by Lake Book Manufacturing

Contents

Part Three: From Moonshot to Mainstream: Alt Meats for All!

Foreword

'm excited to endorse this book from both my professional perspective, as a food security and national security policy thinker, and a more personal one, as a mom.

By day, I direct a program on global food and water security at a national security think tank in Washington, DC. It's through this work that I met Bruce Friedrich and conducted some of the research he describes in the latter part of this book. I decided to investigate alternative proteins, or alt-meat, because of the national security threats associated with animal protein production, which I first noticed during my tenure in the national government, including as part of the US national security community. In the United States, animal agriculture plays an outsized role in the Colorado River Basin water crisis, for example, which affects lives and livelihoods in seven states and two countries. Water stress is threatening energy production, industry, and municipalities across the region, all while a full 55% of the Colorado River's water is used for animal feed—alfalfa and other grasses and corn—to produce meat, milk, and cheese. Further south, in Brazil and neighboring countries, forests are razed to grow other types of animal feed and graze livestock, accelerating biodiversity loss and shrinking the Amazon Rainforest, known as the "lungs of the planet." The loss of biodiversity, including agricultural bio-diversity, undercuts agricultural production itself and threatens economic security and food security for millions of people. Animal agriculture is also closely tied to the threat of zoonotic diseases, which jump from an-imals to humans. Today, 75% of globally emerging pathogens are linked to animals.

The reliance on animals as a principal source of protein also exposes vulnerabilities in food systems. This decade alone, threats have come in the form of cyberattacks, like the attack on infrastructure of JBS, one of the world's top meat producers, in 2021, attributed to Russia; sup-ply-chain disruptions like the ones precipitated by Russia's invasion of Ukraine, which shocked global agriculture markets and pushed the prices of fertilizer and grains (essential inputs for animal agriculture) to record highs in 2022; and pathogens like bird flu, which forced US egg prices to record highs in early 2025, or the New World screwworm, a parasitic fly threatening US livestock, about which federal and state lawmakers are sounding the alarm as I type this. Other risks are less well understood. Policymakers have yet to appreciate the threat of China's dominance of supplies of the vitamins, amino acids, and minerals that fortify US ani-mal feed. Losing access to these micronutrients would increase mortality among livestock in the United States, putting additional pressure on beef prices, which hit an all-time high this summer.

Produced at scale, plant-based proteins and cultivated meats would help mitigate the plethora of national security threats related to ani-mal-protein production: water stress, biodiversity loss, pandemic threats, and the list goes on. As their name implies, alternative meats provide an alternative to conventional meat. And alternatives are becoming more im-portant as the impacts of animal agriculture become clearer, and disrup-tions to food systems become the norm rather than the exception. These days, the only certainty in agriculture is uncertainty, and food systems are more resilient with diversity of choice.

Foreword

Outside work, I'm a mother to my toddler son and two teenage step-sons, whose size I've come to measure not by their age but by their collective length: our house is home to nearly 16 feet of boy and counting. Their favorite food at restaurants is hamburgers, but their favorite food at home is "the pasta," which we make with plant-based protein. Their school district, one of the nation's largest, also offers plant-based options alongside traditional animal proteins. When given a theoretical choice between their favorite sandwich made with conventional turkey or turkey grown from cell culture, they'd gladly take a cultivated-turkey sandwich—"as long as it tastes good."

Bruce's book is an impressive *tour d'horizon* of the need for alternatives to conventional animal protein and how we can make alt-meats—with taste and price parity—a reality. However, his book is not a condemnation of all animal agriculture. As he says from the first page, Bruce does not prescribe how readers should eat, and neither do I. Even with the widespread availability of alt-meats, animal proteins will always remain relevant as a source of incomes for many families, as symbolically significant among many cultures, and as a critical source of iron and micronutrients for women of reproductive age, for whom global anemia levels are a serious and rising challenge.

In *Meat*, Bruce Friedrich balances current realities on the one hand, with a future of possibilities on the other. While many people can't envision food systems that are markedly different from today's, the truth is that food systems are always changing. My sons, like many young people, are already eating alt-meats in their diets and would likely welcome more opportunities to eat cheaper and tastier ones. This book explains the imperative to transform our food systems, and lays out a game plan to get us there.

The time for this book has come, as the discourse about alt-meats is reaching a fever pitch in some corners of the country. Bruce right-sizes the most heated rhetoric about alt-meats, describing a future in which alt-meat is simply called "meat," and the animal protein industry's impacts on food security, water security, and national security are greatly diminished. *Meat* is as important as it is enjoyable. I hope you'll be as inspired to help create an alt-meat future as I am.

Caitlin Welsh
Director, Global Food and Water Security Program
CSIS | Center for Strategic and International Studies
Washington, DC
August 2025

The Next Global Agricultural Revolution

L et me get one thing out of the way right now: I'm not here to tell anyone what to eat.

You won't find vegetarian or vegan recipes in this book, and you won't find a single sentence attempting to convince you to eat differently. This book isn't about policing your plate. Put another way, this book is not about meat at the micro level: what you choose to eat. This book is about meat at the macro level: how meat is made, and how it can be made better.

Here's the question: What if we could make the meat that so many people love, make it just as delicious, and produce it more efficiently, so that it costs less? And what if it could create significant economic growth, improve food security, generate good jobs, and mitigate some of the adverse consequences of modern meat production?

I think we can, and *that's* the topic of this book.

First up, we'll consider plant-based meat. I'm not talking about the veggie burgers and veggie dogs you may have had in years past; I'm talking about making meat from plants that is indistinguishable from animal-based meat: the flavor, the texture, how it cooks, all of it.

Many readers will be dubious about that sentence. You might be thinking, "Veggie dogs suck." Or, "Meat comes from animals; what are you even talking about?"

Sure, most veggie dogs are currently made for vegetarians, and most people who enjoy meat don't enjoy them. But think about this: Until very recently, cell phones were big, bulky things that cost a fortune. And finding a signal often felt like some kind of cruel scavenger hunt. "Can you hear me now?"

Until the late 1990s, most phones had cords that plugged into a wall, and most cameras required film. Smartphones didn't hit 1% penetration until around 2003 in the United States and a few years later globally. It took time and quite a bit of innovation, but now, your phone isn't just your phone, it's also your camera. And your music player. And your GPS. And . . .

I believe that in the not-too-distant future, there will be a variety of plant-based meats that are indistinguishable from conventional animal meat; they'll be made not for vegetarians but for everyone. Once that happens, sales will increase, prices will fall, and market share for plant-based meat will grow aggressively.

Next up, we'll consider cultivated meat: Similar to how we can grow a plant from a seed or cutting, we can grow real animal meat from a small sample of animal muscle or fat, cultivating actual animal meat in a cultivator that looks like a fermentation tank for beer. And while we have some work to do before it can achieve any kind of scale, scientists are making progress on this technology all over the world. And to be clear, this is the same animal meat you're eating now; it's just produced differently.

Cultivated meat is sometimes referred to as "lab-grown," but this is a misnomer. While all processed food starts in a food lab, no one refers to "lab-grown" Corn Flakes or "lab-grown" Twizzlers. Cultivated meat is produced in clean facilities that look a lot like beer breweries. Besides, conventional meat is studied in labs far more than cultivated meat: There are meat sciences departments and meat laboratories at every university on the planet that teach about agriculture, and they have been there for decades. So, 99.9% of current meat labs are studying conventional meat, not cultivated meat.

Together, plant-based and cultivated meat are often referred to as alternative meats, alt meats, alternative proteins, or alt proteins. If the word proteins is used, that generally encompasses dairy and egg alternatives as well.

This is a book about meat, so I won't be talking much about dairy and eggs. Replicating meat presents sufficiently robust scientific (and other) challenges, and those challenges are different from the efforts to replicate dairy and eggs. Plus, while dairy and eggs are enjoyed by most people, there is not the same level of absolute devotion you'll find when talking about steak, bacon, and burgers.

That said, the central theme of the book—that the inefficiencies of conventional meat production create both significant costs and significant opportunities—applies to dairy and eggs as well.

Second, the entire point of this book is that in the future, what are now alternatives to conventional meat will become mainstream, i.e., no longer alternative. So, while I will be referring to plant-based and cultivated meat as alternative or alt meats throughout the book—and so they are, *right now*—my expectation is that in the future, they will be just "meat."

Meat without the need for live animals may sound like science fiction, but consider: For millions of years, ice came exclusively from natural bodies of water, light and heat came exclusively from the sun or burning things, and there was only one way for a woman to get pregnant. Now, "artificial ice" (ice made in a freezer) is just ice, the world's cities are lit up all night long by artificial light, microwaves and convection ovens heat our food, and most of us have friends and family who have enriched their lives tremendously through the wonders of in vitro fertilization. None of this was true just 125 years ago—that's less than a historical finger snap.

There's a third food technology that I would discuss more if this book were intended for scientific audiences: fermentation. Biomass fermentation

3

and plant-based meat present the same basic opportunities and challenges, so I will discuss them interchangeably.*

Precision fermentation creates the ingredient that gives the Impossible burger its meaty flavor and the rennet in cheese, which used to come from calf stomachs but is now synthesized. It's also used to create insulin, which used to come from cattle and pigs. This isn't a science text or a policy brief, and the narrative isn't clarified by a detailed discussion of fermentation. Everything you could possibly want to know about fermentation is available on the Good Food Institute (GFI) website.

But wait—I skipped a step: What's the point of plant-based and cultivated meat in a world that has bean burritos and falafel and tofu and limitless plant-based entrees? I often hear meat lovers express an annoyed bemusement at the idea of veggie burgers: "Don't want to eat meat from animals? Then don't! But stop with the 'pretend meat,'" they'll laugh. "It's not fooling anyone!"

That's what I used to think, too, but I've come to understand that the vast majority of people really love meat—hence the subtitle of this book: It's humanity's favorite food. Plus, meat is deeply rooted in most cultures and is the centerpiece at many social gatherings.

Here's what that means: The average American eats more than 220 lbs. of meat per year, and around the world, people largely eat as much meat as they can afford; one of the clearest trendlines in economics is the correlation between a country's wealth and its meat consumption.[1] Since global wealth has been rising for decades, so has global meat consumption. And that's true everywhere, including India, where meat consumption is three times where it was a quarter-century ago.

I lived my formative years in Oklahoma, and I spent much of every summer and every Christmas holiday in Minnesota, where my parents

---

* The most famous plant-based meat company in the UK is Quorn. The thing is: Quorn doesn't use plants; it uses mycoprotein, which is produced using biomass fermentation. Unless you're a scientist, the difference is not important. The main thing: Plants make their own food; mycoprotein does not.

grew up and all four of my grandparents lived. Oklahoma is the land of cattle and steak houses, and Minnesota is the land of fishing, hunting, and turkey and pig farms. It's also the home of Hormel (the makers of Spam) and Cargill, America's second-largest meat company—and my maternal grandfather's employer for his entire professional life.

Growing up, I ate meat for lunch and dinner every day, and my favorite foods were McDonald's Big Macs and KFC chicken. As far as I can remember, those were the favorite foods of all my friends too. I can't recall meeting any vegetarians in Oklahoma or Minnesota, and the first time someone told me that she didn't eat meat—my first year of college—she may as well have told me she didn't breathe oxygen. I thought maybe she had some kind of disease. Was she allergic? It felt impolite to ask.

Truth is, not a lot has changed. The polling firm YouGov tracks the foods Americans like best. In 2024, nine of the top ten center-of-plate items were meat: hamburgers, cheeseburgers, fried chicken, steak, turkey, ribs, cheesesteak, roast beef, and chicken wings. The one vegetarian item was the grilled cheese sandwich. If you filter for millennials, roast beef drops out of the top ten, replaced by chicken nuggets.[2]

I don't think there's much chance that humanity's love of meat is going to change any time soon; it appears to be biological, as I'll discuss in chapter 5. But we can give people the meat they love, just produced differently: plant-based and cultivated meat that is indistinguishable from conventional animal meat—but safer and healthier.

In part one, I'll answer the question: "Make meat better? What does that even mean?" We'll discuss the inefficiencies of conventional meat production and how that leads to hunger and malnutrition in developing economies, environmental harms, including climate change and deforestation, and two of our world's most concerning global health scourges: antibiotic resistance and pandemic risk. If you know all this or just want to learn more about alt meats, feel free to skip straight to part two.

In part two, I'll suggest that humanity's best response to the steady upward trajectory of meat consumption globally is to make it differently:

plant-based and cultivated meat that offers an identical eating experience but for a lower price. For at least 50 years, environmentalists, global health experts, and animal advocates have been trying to convince the world to eat less meat. Yet consumption is as high as it's been in world history. Every year we set a new record, and absent a global recession or another Covid-like shock, we'll be setting records every year at least through 2050; that's as long as anyone's forecasting.

While recent scientific progress regarding both plant-based and cultivated meat offers cause for optimism, neither endeavor is scientifically simple. The difficulty of these endeavors has led to struggles for plant-based and cultivated meat startups as well as pessimism in some quarters about the entire idea of making meat in these new and better ways.

The observations that led to the pessimism are these: Plant-based meat is too expensive, and most of it doesn't taste very much like animal meat. Cultivated meat is not yet widely available and costs are way too high, despite more than three billion dollars in sector-wide investment. Three. Billion. Dollars. And hey, Richard Branson and Bill Gates have invested; if it's not solved yet, it never will be.

The observations are accurate, but the conclusions forget how innovation happens, how recent both endeavors are, and how little $3 billion is when spread across more than 150 companies.

In part three, I will add that context and build from it: The struggles of plant-based and cultivated meat are typical in the annals of innovation. We'll consider other technologies that struggled in their early years and only reached mainstream adoption after critical improvements to early iterations: from cars to airplanes to widespread computer use to online shopping to large language models that can write poetry and outperform doctors in diagnosing disease.

We'll also explore the three-legged stool of scientific innovation, which involves cooperation across science, industry, and government. We'll dive into a Center for Strategic and International Studies report on the economic competitiveness and food and water security benefits of alternative meats, and we'll explore some of the excellent plant-based and cultivated meat scientific work that governments are supporting now. We'll close part three by diving into the question that all of us at the GFI hear more than any other: How quickly can we expect to find taste- and price-competitive alternative meats in our local grocery stores and fast-food restaurants? Put another way: How soon until we can drop the "alternative" from alternative meats?

I'll close the book with some unsolicited career advice for anyone who is convinced that alt meats offer tremendous value to the world and who would like to help make it happen as soon as possible.

My Journey: From Serving Soup to Remaking Meat

But first, a little bit about me: In 1984, I was growing up in a suburb of Oklahoma City and was confirmed into the Evangelical Lutheran Church of America. This was during the height of the famine in Ethiopia, and the pastor of my congregation, David Klumpp, was devastated by the suffering there. So my two years of confirmation classes focused on global hunger and how Christians should respond.

Matthew 25 provided the answer: "As you did it to the least of these, you did it to me." That and the story of the good Samaritan, which teaches that all of humanity is our neighbor. Our moral obligation as Christians, Pastor Klumpp told us, was to help whoever in the world was suffering the most.

That early exposure to injustice, my own privilege, and what it means to act ethically in a world of want set me on a lifelong path of trying to make the world better.

At Grinnell College in Iowa, I joined a campus group called Poverty Action Now! and volunteered in a Des Moines homeless shelter run by a Catholic priest named Frank Cordaro, who remains a dear friend. I also read *Diet for a Small Planet* by Frances Moore Lappé, which argues that feeding crops to animals for meat is an inefficient use of food and land. In a global economy, that inefficiency leads to hunger and malnutrition.

That book changed my diet, inspired me to major in economics, and led me to write my senior thesis on how the global economy has created an agricultural system that entrenches extreme poverty.

After college, I ran a Catholic soup kitchen and homeless shelter for families in Washington, D.C., for six years (and converted to Catholicism), taught through Teach For America in inner city Baltimore for two years (teacher of the year for my school, thank you very much!), and earned a teaching degree from Johns Hopkins and a law degree from Georgetown. I also worked to improve laws and regulations related to farm animal treatment and ran public education campaigns to (try to) inspire people to eat less meat.

I thoroughly enjoyed running the shelter for families and soup kitchen, loved teaching even more, and enjoyed my work to improve farm animal welfare and inspire people to change their diets. But I also wondered constantly, "Is this the most effective thing I could be doing to make the world better?"

Inspired by a handful of food systems pioneers who you will meet in chapters 6 and 7, I came to believe that there is a promising option that—if it works—will do tremendous good for the world: plant-based and cultivated meat that is designed to be just as delicious and affordable as conventional meat.

Alternative meats are similar to renewable energy and electric vehicles. The world is going to consume more energy and drive more miles as it gets wealthier. It's also going to eat more meat. That's why we need renewable energy, EVs, and alternative meats. None of them is a silver bullet, of course; there are no silver bullets in food or energy. But they are essential.

Science and innovation have improved solar and EV technologies, and scaling has brought down costs. The science of plant-based and cultivated meat is extremely promising. With scale, prices can come down, offering consumers the meat they love, but without all the adverse impacts for our environment and global health, and without the contribution to hunger, malnutrition, and animal suffering.

That vision inspired me to launch GFI in 2016 with a focus on accelerating the science of plant-based and cultivated meat (the plurality of GFI's team members around the world are scientists). We do that mostly by working with governments and the food and meat industries to accelerate the trajectory of these food innovations.

Bonus: Scaling up plant-based and cultivated meat production can also create significant economic growth, improve food security, generate good jobs, and open up opportunities to transform the global farm economy in ways that improve farmer livelihoods. Of course, none of this is self-executing; it will require hard work.

This book is my invitation to you, to join me for an abridged version of the journey I've been on. I hope you'll come away from these pages feeling as energized as I am about what's possible, and perhaps also inspired to help make it happen.

Welcome to the next agricultural revolution.

PART ONE

Feeding the Future

The Case for Alternative Meats

CHAPTER ONE

Feeding the Hungry

Investing in alternative proteins now could help address malnutrition, a key driver of which is a lack of access to affordable, high-quality proteins.

— Innovation Commission:
Climate Change, Food Security, Agriculture[1]

The Food Waste Built into Animal Physiology

In my 2017 TEDx talk, I walked onto the stage and asked: "Who likes pasta?" All hands went up, as they always do. I then handed one person a plate of pasta from a local Italian restaurant and tossed another eight plates into a trash can. "I'm sorry," I said. "I only have nine plates, and I have to throw the other eight away." The audience gasped.[2]

Then I asked, "Who here would cook nine servings of pasta and throw eight away?" No one raised their hand.

What was I thinking, tossing eight plates of delicious pasta in the trash? I was making a point: It takes about nine calories of crops to produce one calorie of chicken—the most efficient animal at converting feed into meat. The other eight calories go toward keeping the animal alive or are lost in feathers, bones, and other inedible parts.*

Feed requirements (calories) to produce 100 calories of meat	
Pork	1,000 (900 calories lost)
Chicken	910 (810 calories lost)
Farmed Fish	833 (733 calories lost)
Cultivated Meat	300 (200 calories lost)
Plant-based Meat	100 (no calories lost)

The multipliers for pigs, cattle, and most farmed fish are even worse. Since those animals live longer than chickens—e.g., two years for a salmon or cow against six weeks for a chicken—they consume even more feed per calorie of meat produced.

When it comes to meat, it takes a staggering amount of food to produce food.

———

When global leaders gathered at the UN Sustainable Development Conference in 2012, they committed to eradicating extreme poverty and hunger by 2030. The goal felt ambitious but within reach. The twentieth century had seen tremendous progress; with extreme poverty falling by about 45 million people per year, it seemed we'd get there with time to spare.[3]

Sadly, that's not where we are now. Each year, the UN Food and

---

* The World Resources Institute points out in *Creating a Sustainable Food Future* that feed conversion ratios are sometimes posited of 3 to 1 or even better. These ratios are misleading for two reasons: First, they are mass ratios, so you're comparing animal feed, which is 10–15% water, to chicken meat, which is 65–70% water. If you compare energy (calories), the best a chicken can do is about 9 to 1. Second, they often use live weight, which includes blood, bones, feathers, and other inedible bits. *Creating a Sustainable Food Future* (World Resources Institute, 2019), https://www.wri.org/research/creating-sustainable-food-future.

Agriculture Organization (FAO) publishes their assessment of global food security and nutrition. In 2024, the agency reported that between 2015 and 2022, the number of moderately or severely food insecure people jumped from 1.6 to 2.3 billion, with severe food insecurity rising from 570 to 870 million.[4] The UN now estimates that we'll miss the "zero hunger" target by more than half a billion of our human sisters and brothers, almost all in developing economies.

I don't know about you, but for me, these kinds of statistics can get pretty mind-numbing. I often think about a Nicholas Kristof column in the *New York Times* about the nature of empathy.[5] We all know that line from Joseph Stalin, that a single death is a tragedy, but a million deaths is a statistic. It turns out that in fundraising, donations double when you focus on one child rather than eight. Donations are also higher for one child as opposed to two. What then to do with hundreds of millions, other than remember that they are all individuals who are as worthy of compassion as members of our immediate family.

Beyond hunger and malnutrition, the report explains that more than one-third of the world's population, almost 3 billion people, could not afford a healthy diet in 2022. That includes about two-thirds of the population of Africa, one-third of Asia, and one-quarter of Latin America.[6] Why is a nutritious diet out of reach for more than a third of the global population? One critical factor identified in the report is food prices, which just keep rising.[7]

Macroeconomics: When Food Becomes Unaffordable

Of course, corruption, conflict, and war contribute to hunger. But they are only a part of the story. Lappé's core insight from 1971 is still true today: A huge share of agricultural exports from poor countries goes to rich countries, much in the form of animal feed. This drives up prices for both food and land, pushing out small-scale farmers and making food unaffordable for the world's poorest inhabitants.

And it's not just cereals that see higher prices: FAO has warned that hundreds of millions of pastoralists and smallholders who rely on livestock for survival are being squeezed out by large-scale agribusinesses that are competing for land and water.[8]

Perhaps the most influential economist on global hunger is Amartya Sen, who won the Nobel Prize for his work on welfare economics. In his landmark book *Poverty and Famines: An Essay on Entitlement and Deprivation*, Sen showed that starvation isn't always caused by food being unavailable; as often, it's caused by food being unaffordable. Of course, politics and weather and conflict all contribute to hunger—largely because they influence affordability by prompting price spikes or exacerbating income inequality. As the FAO report makes clear, the principal driver of food insecurity for billions of people is the too-high cost of food.

For example, in the 1943 Bengal famine, rising urban wages drove up food prices, but rural wages didn't keep pace. Millions of rural workers starved—not because there wasn't enough food, but because they couldn't afford it.[9] Similarly, the 1974 Bangladesh famine was triggered by flooding that spiked food prices and slashed farmer wages. Again, the food was there, but it was too expensive for the people who needed it most.

Fast forward to 2007: UN Special Rapporteur on the Right to Food Jean Ziegler held a press conference in New York to declare biofuel production "a crime against humanity." Why? Because land that could be used to grow food was being used to grow fuel, which drove up land and food prices, leading to increased hunger and malnutrition. A year later, a leaked World Bank report estimated that biofuels had driven up global food prices by as much as 75%, putting fuel tanks in direct competition with the bellies of the world's poor.[10]

What Ziegler missed—or perhaps thought too intractable for comment—is that far more crops are turned into animal feed than are turned into biofuels. At the time, *Guardian* columnist George Monbiot pointed out that nearly eight times as much corn and wheat went to animal feed as to biofuels. Resource economics are complex, so we can't say that feed crops caused exactly eight times the price pressure or hunger impact. It could be a bit more or a bit less.

But no matter what, using land for feed crops causes many times the price pressure and many times the hunger impact of using land for biofuels, which the UN's own hunger envoy called a crime against humanity.[11]

That was 2007, and as meat production has increased, so has the amount of corn, wheat, soy, and other crops that are fed to farm animals. By 2021, animals were consuming 1.1 billion metric tons of cereals in a year.[12] Add in soy—mostly fed to chickens, pigs, and farmed fish—and that total jumps to 1.37 billion metric tons.*

To grasp how staggering that number is, consider the impact of the war in Ukraine on global grain markets. When Russia invaded, wheat exports from Russia and Ukraine—about 48 million metric tons—were disrupted. That's less than one-twentieth of what we feed to farm animals each year. But it was enough to send food prices soaring and to spark fears of famine in vulnerable regions.[13]

The TEDx audience gasped when I tossed eight plates of pasta into a trash can. But 1.37 billion metric tons is the equivalent of 7.6 *trillion* plates of pasta.† It's hard to know what the proper reaction is to that level of food waste. But one thing is for sure: Demand for meat is driving up food prices, and that's exacerbating hunger and malnutrition.[14]

---

* That's 1.1 billion metric tons of cereals and 270 million metric tons of soy. This figure does not include the 232 million metric tons of food (e.g., roots, vegetables, and pulses) that were fed to farm animals, up from 152 million metric tons in 2001, according to *Livestock's Long Shadow*. Henning Steinfeld, Pierre Gerber, Tom Wassenaar, Vincent Castel, Mauricio Rosales, and Cornelis de Haan, *Livestock's Long Shadow: Environmental Issues and Options* (Food and Agriculture Organization of the United Nations, 2006). 232 million metric tons calculation courtesy of Hannah Ritche, using the FAO calculator. "FAOSTAT: Food and Agriculture Data," Food and Agriculture Organization of the United Nations, last updated June 12, 2025, https://www.fao.org/faostat/en/.
† The average plate of pasta weighs about 180 grams. That said, animal feed is very low in water, so it would be more accurate to compare animal feed to the weight of dry pasta, so 90 grams, which would double that 7.6 trillion number.

Alt Proteins: Lower Prices, Less Hunger

Intuitively, one would imagine that if demand for meat is driving up prices, then a shift to alt meats that are produced more efficiently should lower them—but is that intuition correct? Yes, according to three independent analyses: from the University of Chicago; from ClimateWorks Foundation and the UK government; and from the International Institute for Applied Systems Analysis (IIASA).

In 2023, Nobel laureate in economics Michael Kremer launched a commission at the University of Chicago to identify agricultural strategies that address both climate change and hunger. And then most importantly, to warrant the commission's endorsement, a strategy must create an economic incentive for further adoption. That's both because governments are looking for solutions that bolster their economies and also so that once implemented, the interventions can sustain themselves.

Scientists from GFI, along with bioscience entrepreneur Grant Gordon and his colleagues at his Nairobi-based startup Essential, submitted briefs laying out the case for alternative proteins. We met with Kremer's team, and they dove into our supporting documents and did their own deep research. Ten months later, they reported that alternative proteins could mitigate climate change, help countries adapt to the impact of climate change on agriculture, and reduce hunger and malnutrition, corroborating findings from our own analysis.*

Two other analyses reached the same conclusion: McKinsey economists working on behalf of ClimateWorks Foundation and the UK government found that at 50% alt protein adoption, crop prices would fall by 10–12%, whereas on their current trajectory, they would continue to rise. IIASA's analysis found a more modest impact on crop prices, a drop

---

* The seven interventions they ended up supporting are: Improved Weather and Seasonal Forecasts; Training for Rainwater Harvesting; Microbial Fertilizer; Innovations to Reduce Livestock Methane Emissions; Digital Agriculture; Climate-Resilient Social Protection; Alternative Proteins. *Recommendations for COP28.*

of 5%; their analysis also found that meat and dairy prices would fall by three times as much—so a shift toward alt proteins would also make meat and dairy more affordable. Both analyses state explicitly that increased adoption of alternative proteins will lead to decreased hunger and malnutrition.[15]

Microeconomics: Alternative Proteins to Feed the Hungry

For many years, I have believed that developed countries should eat less meat, so that developing countries could have any food at all. The argument is straight out of *Diet for a Small Planet*: Rich countries eat too much meat and fish, which increases prices for land and food. It also decimates fish populations, which cripples subsistence fishing communities.

We need to eat less, which could free up land for pastoralists and smallholder farmers, allow fish populations to bounce back, and lower food prices, leading to lower rates of hunger and malnutrition. That logic helped turn me vegetarian and fueled much of my early advocacy.

That's all correct, as discussed above, but over the past few years, I've been convinced that alternative proteins could also have a strong role to play in direct feeding programs.

Consider the work of Sarah LaHaye, who works at the Global Alliance for Improved Nutrition, an NGO that is focused on eradicating hunger and malnutrition. Previously, Sarah spent eight years at One Acre Fund, an organization supporting smallholder farmers in Eastern and Southern Africa with tools, training, and commercial support. Their programs increase farmer incomes by an average of 40% and provide meaningful partnerships to millions of the world's poorest people.

While at One Acre Fund, Sarah launched a program in Rwanda to help farmers grow crops that could be turned into texturized vegetable protein (TVP) products that, as she explained to me, "fulfill the universal

19

human desire for meat, but at a much lower cost—something rural Rwandans could afford." During a COP29 panel that GFI co-hosted with FAO and CGIAR, she described the project's twin goals: The crops had to be nutritious, climate-resilient, and good for soil health, and the final product had to be desirable and marketable.[16]

Partnering with scientists from Wageningen University in the Netherlands, the One Acre Fund team used indigenous crops to create a meat-like product. The results surpassed their hopes, as Sarah explained during the panel: "The biggest takeaway was the successful perception of eating a meal with animal meat—something families really aspire to."

When I spoke with Sarah after the panel, she shared more: In smallholder settings, meat production can pose serious food safety risks due to lack of refrigeration and clean handling infrastructure. Livestock can also introduce pathogens and contaminate shared living spaces. TVP, on the other hand, is safer to handle, faster to cook (saving on fuel use and slashing air pollutants), and can be processed to reduce anti-nutrients in beans. The benefits multiply.

Grant Gordon: From Evaluating to Implementing Development Interventions

When Grant Gordon met with Michael Kremer's team to share his view that alternative proteins could be a powerful intervention for climate mitigation, climate resilience, and hunger, he felt confident about his analysis. That's because Grant had spent the previous 15 years in global development focused on creating breakthrough innovations to tackle the most important problems of the poor. While serving as senior director of innovation strategy for the International Rescue Committee, he worked on malnutrition and maternal and neonatal health.

"Two problems stood out," he told me, "which sparked Essential." The first was that proteins are by far the most expensive component of products used to prevent and treat malnutrition. So that's what was screaming out for innovation, in Grant's mind. Second, climate change was ravaging traditional agriculture and leading to hunger and malnutrition. "It was clear that creating low-cost, high-quality, *climate resilient* proteins was urgently needed."

Grant's company, Essential, runs a biomanufacturing plant in Nairobi, Kenya, that is using biomass fermentation to create proteins that can be integrated into a wide variety of products that prevent and treat malnutrition. They're expecting to produce more than 3,500 metric tons of high-quality protein for less than $2 per kilogram by 2030. Grant's vision is that his proof of concept will lead to the scaling of this technology as a cost effective and impactful hunger intervention.

The Gates Foundation sees the promise too. It provided a $4.76 million grant to the Chicago-based alt protein company Nature's Fynd to develop what is essentially the alt proteins version of a bread maker, but for people in low- and middle-income countries. If successful, smallholders would be able to use agricultural waste streams to make high-quality protein, for both personal use and sale.[17]

That exact same concept is a motivating factor to one of the pioneers of cultivated meat research, tissue engineering professor Marianne Ellis from the University of Bath in the UK. She envisions cultivated meat facilities in the poorest regions of the world, both as a strong climate resilience intervention and also for global development: In her mind's eye, she pictures buildings "the size of six shipping containers stacked three layers high, covered in solar panels." They could not only provide high-quality nutrition; they could also create a robust local economy, center community activity, and become an engine for "schools and education, job prospects, better health and quality of life."[18]

Alt Proteins for Feeding Programs

Kremer's team noted that alternative proteins have the potential to be a cost-effective intervention for hunger and malnutrition. Because these proteins are high-quality and significantly cheaper than conventional meat and dairy, the commission posited, alternative proteins could be a lifeline for countless families in the poorest parts of the world.

Even beyond the kinds of examples I've discussed above, Kremer's team argued that alt proteins have the potential to serve as a primary protein for feeding programs run by UNICEF and the World Food Program, which rely on fortified pastes that are simple but life-saving. Kremer's team noted that due to lack of funds, half to three-quarters of severely malnourished children don't receive these foods. Because they are much less expensive, alternative proteins could significantly increase the number of hungry people served.

On the same COP29 panel where Sarah LaHaye shared her story, University of Chicago policy professor Amir Jina—also a contributor to the work of the Innovation Commission—drove the point home: "The single biggest price component in those foods is milk powder. Shifting that to an alternative protein could dramatically lower the price . . . anything we can do to extend that supply is critically important." Alternative proteins could make nutritious food cheaper, safer, and more available—a rare triple win in global nutrition.

Alt Proteins and Fortification

Reading through the Kremer commission policy brief and other evidence, one finding continued to befuddle me: Animal-source proteins were consistently much more expensive—often more than twice the cost—relative to their animal-free comparators, yet often did not deliver better nutrition outcomes.[19] Why do feeding programs continue to rely on them?

In a word, micronutrients. Soy and mycoprotein already have just as much bioavailable protein as animal-source foods, but they are lacking in critical micronutrients.

This feels to me like a problem with an obvious solution: We already add iron, zinc, and folic acid to flour and vitamin A to oils and sugar. Why not extend this to alternative proteins, fortifying with key nutrients of concern such as iron, zinc, and vitamin B12? Better still, we could invest in biofortification: breeding staple crops to contain these critical nutrients. These enhanced crops—such as iron-fortified soy—could then become the raw ingredients in alternative protein products, making them even more nutritious.

Grant told me that the cost to produce plant protein or mycoprotein fortified with the key nutrients of concern would likely be much smaller compared to the costs of animal production. It's also much simpler, which would make execution far more likely to succeed, a critical concern in rural areas especially. Scaling up animal production affordably would require mimicking the same factory-farm production methods used in OECD countries, which would take decades and many billions of dollars. "There are much more efficient ways to meet protein and micronutrient needs in LMICs," Grant explained.

The upshot? Alternative proteins are not only cheaper, but they can also be nutritionally competitive. With the right focus, they can mitigate hunger and malnutrition both through their macroeconomic effect on land prices, food prices, and fishing stocks, and also as a direct way of nourishing some of the poorest members of the human family.

Climate Resilience: Food Systems for a Warming World

One final point: More and more, the global community that works on climate change is shifting some of its attention from mitigation to adaptation, in recognition of the fact that climate change is here, and we'd better figure out how we live with the changes that have already been locked in. As usual, it's the global poor who are most adversely affected by the impact of climate change, despite having done the least to contribute to it.

As global temperatures continue to rise, we'll see diminishing viable farmland, falling crop yields, more frequent and more severe droughts and flooding, which will destroy crops and kill livestock, and rising water temperatures and acidification that will create uninhabitable conditions for fish and other aquatic life.[20] For example, a team of sixteen economists found that climate change could lead to global crop yield losses of 24–45% by the turn of the century, including losses of 19–54% in South America, 28–45% in sub-Saharan Africa, and 38–47% in North Africa.*

In addition to their climate mitigation benefits, alternative proteins can be a critical climate adaptation intervention, because they require less land and less water than animal agriculture, and they're far more adaptable in the face of environmental volatility.

For example, mycoprotein facilities can convert agricultural waste into high-quality protein, and they can do this 365 days a year, 24 hours a day. They can also operate in cities and in other places without much arable land and even in drought zones, regardless of droughts or other weather extremes that can destroy crops and kill animals. Assuming we figure out how to scale the facilities and slash costs, cultivated meat will offer these same benefits, eventually.

We live in a world of extremes: Hundreds of millions of people are hungry, and almost three billion can't afford a healthy diet. But we also have both greater affluence and more overweight people than ever before in human history.

---

* "Estimating Global Impacts to Agriculture from Climate Change Accounting for Adaptation" (June 26, 2025). Available at SSRN: https://ssrn.com/abstract=4222020.

If you're reading this book, I'm guessing you care about food systems. And if you care about food systems, I'm guessing you care about food loss and waste.* Me too. I find it maddening that for every 100 calories of food produced globally, almost a quarter of those calories are lost or wasted. Remarkably, that percentage is getting worse as our food system becomes more global, and it's expected to double by 2050.[21]

When I tossed eight plates of pasta into the trash, the crowd showed its dismay. That instinct—that food shouldn't be wasted—is right. But think about this: For every 100 calories of chicken we eat, another 800 calories are spent keeping the animal alive or producing nonedible parts of the bird. That's a sort of food waste that's baked into the process of cycling crops through animals for meat, and it's responsible for 30 times as many lost calories as direct food loss and waste.

The inefficiency of feeding crops to animals so that we can eat animals is what piqued my interest in global agriculture 40 years ago, and it's also half of the founding challenge of GFI, which we framed this way: How are we going to grow enough food to feed almost 10 billion people in 2050, without burning the planet to a crisp?

Alt meats can help: The more we shift toward plant-based and culti-vated meat, the more we can free up land and lower food prices, allowing more and more of the world's most vulnerable to eat.

---

* According to the World Resources Institute, "Food loss refers to loss at or near the farm and in the supply chain, for example, during harvesting, storage or transport. Food waste occurs at the retail level, in hospitality and in households." Liz Goodwin, "The Global Benefits of Reducing Food Loss and Waste, and How to Do It," World Resources Institute, April 20, 2023, https://www.wri.org /insights/reducing-food-loss-and-food-waste.

CHAPTER TWO

Slashing Emissions and Saving Rainforests

Alternative proteins will produce the same total calories as [conventional animal] proteins but do so using 640 million fewer hectares, thus freeing land for nature and making both biodiversity and deforestation targets easier to meet . . . Soybean production, which is heavily used for feed production and is often linked to deforestation of the Amazon rainforest, would halve by 2050, reducing both deforestation risk and pastureland, thereby allowing for the restoration of natural ecosystems, including biodiverse and carbon-dense forest area.

— Climateworks Foundation and the UK Foreign, Commonwealth, and Development Office[1]

Alt Meats: The Electric Vehicles of Food

So Bill Gates and a rocket scientist walk into a bar . . . Well, not exactly. But a few years back, Bill did have lunch with former NASA engineer Mark

Rober, host of one of YouTube's most popular science channels.[2] In a video that's been viewed tens of millions of times, Gates and Rober bond over their shared love of meat—and their discomfort with its environmental cost.

For a long time, Gates assumed that of all the toughest climate challenges, meat production would be the hardest to solve. That's because he believes the key to solving climate change is replacing emissions-intensive products with innovative, low-emission alternatives. And until very recently, there was no plausible substitute for conventional meat.

Vegetarians and meat reducers may find that hard to hear: Why not just eat something else? But Gates, looking at decades of meat-reduction campaigns that had not led to decreased global meat consumption, as well as his own love of burgers, sees this as another version of the "small is beautiful" philosophy that encourages people to consume less energy as a solution to the climate impact of fossil fuels: admirable, but extremely unlikely to work on a global scale.

With steadily rising global meat demand that stretches back to forever and no indication that consumers anywhere had much appetite for changing their diets, Gates wasn't feeling particularly optimistic about reducing meat's climate footprint. Then came plant-based and cultivated meat—agriculture's answer to renewable energy and electric vehicles (EVs), in that they are market-based solutions that focus on supply instead of demand.

Renewable energy, electric vehicles, and alternative meats are focused on giving consumers everything they like about energy from fossil fuels, gas-powered cars, and conventional meat, but at a lower price. Instead of changing the human desire for convenience that fuels energy consumption and personal transportation, let's change the way those needs are met. Instead of changing the innate human desire for meat (more on this in chapter 5), let's change the way that meat is produced.

The strategy only works if the products win in the marketplace: For that to happen, we'll need to bring down prices and improve functionality—what Gates calls eliminating "the green premium," the current extra cost in functionality or money required to opt for the greener product.[3]

Early electric vehicles faced multiple premiums: higher prices, limited battery range, and a lack of charging infrastructure. And until recently, solar energy was a lot more expensive than energy from fossil fuels. But with innovation and scale, costs for both technologies fell, battery range improved, and chargers proliferated. In response, EV and solar adoption have risen quickly, as I'll discuss more in chapter 8.

To be clear, both functionality and price are critical: Most consumers weren't going to buy a price-competitive EV if the battery range was just 80 miles or there was no charging infrastructure. And they weren't going to pay more for an EV, even if it had a reasonable battery range and ample places to charge up.

Same with alternative meats: Consumers love meat because it's delicious and affordable; they are not wedded to its current method of production. There is extensive research to indicate that plant-based and cultivated meats that can match both the deliciousness and price of conventional meats can arrest and reverse meat's global upward trajectory (covered in chapters 6 and 7). As a quick aside, when I use the word taste, I'm including texture, aroma, and the whole eating experience, what food scientists call "organoleptic qualities."

Gates is backing that bet: He was an early investor in plant-based pioneers Beyond Meat and Impossible Foods and one of the first backers of the world's first cultivated meat company, Upside Foods. He discusses his enthusiasm for these technologies in his accessible and optimistic book, *How to Avoid a Climate Disaster.*

Code Red for Humanity: The UN Secretary General Declares a Climate Emergency

The most authoritative report on climate change is the Intergovernmental Panel on Climate Change's (IPCC) Sixth Assessment Report (AR6), released in four parts between late 2021 and mid-2023. The first release detailed the scientific impacts of human-caused emissions. Its conclusion was stark: Climate change is driving record-breaking heat waves, devastating floods, prolonged droughts, intense wildfires, and stronger hurricanes. It's long past time for aggressive action to mitigate worsening impacts.

In response, UN Secretary General António Guterres didn't mince words. He called the report "code red for humanity," warning that emissions from fossil fuels are "choking our planet and putting billions of people at immediate risk." Global heating, he said, is affecting every region on Earth, with many changes irreversible.[4] The world's most vulnerable are in the greatest jeopardy.

Guterres continued to sound the alarm. In a 2023 press conference, he declared: "The era of global warming has ended; the era of global boiling has arrived . . . Climate change is here. It is terrifying. And it is just the beginning."[5] He painted a grim picture: "The air is unbreathable. The heat is unbearable . . . Children swept away by monsoon rains; families running from the flames; workers collapsing in scorching heat."

Later that year, he turned (even more) apocalyptic: "Humanity has opened the gates of hell. Horrendous heat is having horrendous effects. Distraught farmers watching crops carried away by floods, sweltering temperatures spawning disease, and thousands fleeing in fear as historic fires rage."[6]

Guterres also laid out an urgent blueprint for action: Stop oil and gas expansion, halt new coal development, end fossil fuel financing, and shift energy investments to renewables. He demanded detailed transition plans from financial institutions and energy companies alike: no more greenwashing, no more deception.

The Secretary General's admonitions were impressively human; the man is clearly outraged by our warming planet and its crushing impact on the global poor. And he is personally committed to doing all in his power to help. That includes naming names.

The Secretary General's Climate Blind Spot

But Mr. Guterres has consistently neglected a piece of the puzzle without which climate targets are scientifically impossible: animal agriculture. I reviewed about a dozen of his climate speeches and statements, and I found nothing at all on the issue.

The oversight is remarkable, because every model that keeps global warming to below two degrees Celsius relative to pre-industrial levels (the *less aggressive* globally agreed target under the Paris Agreement) requires both a significant reduction in emissions from farm animals and the sequestration of billions of metric tons of carbon, including through nature-based interventions like reforestation. Neither of these outcomes is possible unless conventional meat production declines.

The World Resources Institute, working with the World Bank and the UN's environment and development agencies, produced a sweeping analysis of food systems and climate. Their conclusion? To keep climate change within Paris climate agreement limits, agricultural emissions must fall to four billion metric tons by the year 2050.* On our current trajectory, the analysts predict emissions totaling almost four times that number: 15 billion metric tons. Hitting WRI's target would require a major decrease in conventional meat production.[7]

The authors also note that meeting Paris targets will require "reforesting hundreds of millions of hectares of liberated agricultural land." But WRI warns that if we stay on our current meat consumption trajectory, there will be no land to reforest. In fact, all that extra meat will require either a massive increase in land productivity or an additional 3.3 billion hectares of land for grazing and feed crops. Just to give you an idea of how

---

* When I refer to carbon or CO2 emissions with precise numbers, these are carbon dioxide equivalent numbers (CO2eq), using the 100 year multiplier for methane. As discussed below, over 20 years, the contribution of animal agriculture to warming is even greater.

much land that is: Three billion hectares is about the size of China and India combined. Times two. Plus Indonesia.

WRI warns that devoting that much additional land to agriculture would require wiping out the world's remaining forests and savannas.

The Secretary General has used another metaphor to capture the climate emergency that is worth noting: "We are on a highway to climate hell with our foot on the accelerator," he declared in 2022. The thing is, all his recommendations combined wouldn't take humanity's foot off the accelerator; it would simply shift our collective foot back about two-thirds of the way.

Important? Of course. But is this code red for humanity, or isn't it? Even if we follow the Secretary General's direction to the letter, we will still blow past climate targets and usher in his nightmare scenarios.

Make no mistake: Global meat consumption has been going up for as long as records have been kept, and all indications are that the upward trajectory will continue—every year, the world sets a new record for both per capita and total meat consumption.[8]

Until we figure out how to bend that curve, there's no plausible scenario that allows us to achieve climate or biodiversity targets. We need all of the energy transition work, obviously, but we also need protein transition. Both are required to take our foot off the accelerator.

Meat's Heavy Climate Footprint

Let's take a closer look at why all the climate models require that the world eat less meat to meet climate targets. We opened chapter 1 by detailing the basic inefficiency of meat: It takes about nine calories of crops to produce one calorie of chicken, and even more for pork, beef, and farmed fish. That inefficiency magnifies every environmental impact: land and water use, water and air pollution, biodiversity loss and climate change.

The chain of inefficiency doesn't stop with feed. Once the crops are grown, they must be shipped to feed mills, processed into animal feed, and transported to industrial farms. Then the animals must be raised, shipped to slaughterhouses, processed, and the meat refrigerated and shipped once more to warehouses and stores. Every step releases emissions. Every step compounds environmental damage.

In 2006, the UN Food and Agriculture Organization (FAO) assessed the environmental consequences of all these inefficiencies and extra stages of production.[9] The effort included agricultural scientists from their own organization and the World Bank, European Union, the International Fund for Agricultural Development, and other government agencies across Europe and around the world; it manifested in a 416-page report, *Livestock's Long Shadow: Environmental Issues and Options.*

The report pulled no punches, concluding: "The livestock sector emerges as one of the top two or three most significant contributors to the most serious environmental problems, at every scale from local to global." For example, meat production is a top three cause of land degradation, climate change, air and water pollution, water scarcity, and biodiversity loss.

That was 2006, when meat production globally totaled about 266 million metric tons. Less than 20 years later, meat production had risen to 370 million metric tons (not including seafood). All predictions are that it will keep rising.

In 2024, the World Bank estimated that livestock accounted for about one-fifth of global emissions, including the plurality of methane (cattle digestion alone creates more methane than oil and gas production combined).[10] That number could be seen as conservative, because it uses methane's warming impact over a 100-year period. Methane's impact over 20 years—a timeframe considered particularly crucial by many climate advocates—is roughly three times greater.

Meat and Climate: The Secretary General is Not an Outlier

I shouldn't be so hard on Mr. Guterres. In fact, most detailed climate reports have little in the way of tractable strategies to mitigate agriculture's emissions, and most climate experts and advocates focus overwhelmingly (and in many cases exclusively) on fossil fuels, just like the Secretary General.

Here's one of the more remarkable examples: Every year, tens of thousands of climate diplomats and representatives of NGOs gather at the Conference of the Parties (COP) to negotiate plans focused on meeting climate targets; at COP21, which took place in Paris in 2015, 194 nations and the European Union agreed to the Paris Agreement, which has guided global climate conversations and policy ever since. For the first 27 years of climate conferences—until 2023—agriculture was not on the agenda to any meaningful degree.[11] Even today, discussions about agriculture are largely devoid of the kinds of scalable and market-based solutions that are central to energy discussions.

This might not be a big deal if progress on energy transition had emissions mitigation covered. But it doesn't: First, emissions from fossil fuels continue to rise as coal, oil, and gas production all hit new records year after year. Second, climate pledges from the G20 nations, the countries that create most emissions, continue to fall dramatically short of Paris Agreement goals. And third, even those too-weak pledges aren't being met.[12]

Another important point: Even if all the world's electricity were produced using renewable energy, that would leave about 45% of emissions unaddressed. That's because a quarter of global climate emissions come from methane and nitrous oxide, and a fifth of emissions come from either land use change (e.g., deforestation) or the so-called hard-to-abate sectors (e.g., shipping, aviation, iron and steel, chemicals and petrochemicals, and heavy-duty trucks).[13]

The top cause of methane emissions, nitrous oxide emissions, and deforestation? Animal agriculture. So far at least, all these sectors are breaking emissions records every single year, and although there was a tiny Covid-inspired emissions dip, the world was setting records again by 2023, and emissions trendlines look like Covid-19 never happened.[14][15]

Ignoring meat isn't just a missed opportunity. It guarantees that we will not meet climate targets.

Alternative Proteins: One Essential Tool in Our Emissions-Mitigation Toolkit

It's not fair to say that climate world is totally ignoring meat's climate impact. Although energy transition is the focus of most climate reports, academic climate papers, and climate advocacy efforts, attention to meat production is real and growing. The three dominant strategies that climate advocates tend to focus on are changes to cattle production; shifting agricultural production to regenerative methods; and promoting plant-rich diets.

These efforts are important, and I'm very firmly in the "all of the above" camp where emissions mitigation is concerned. But it's also true that even best-case scenarios for all these interventions combined won't cut agricultural emissions enough for us to come anywhere near hitting climate targets. Recall that WRI's analysis says that we need to decrease agricultural emissions to four billion metric tons by 2050 to meet climate goals (that's 11 billion fewer metric tons than we can expect on our current trajectory), and we need to re-forest hundreds of millions of hectares of liberated agricultural land.

The success of alternative meats will be essential to both halves of WRI's prescription.

Think about it this way: There's a lot of critical work happening to improve energy efficiency for everything from buildings to light bulbs; to reduce energy consumption; to reduce the climate impact of air travel and long-haul shipping; to create more walkable and bikeable cities; to improve public transit; and so much more. That's all important, of course. But we also need renewable energy if we're going to meet climate goals.

Methane Reduction: Mitigating Cow Burps

Take methane reduction from cattle. Several promising interventions are being explored, such as adding seaweed or chemical additives to cattle feed, improving the digestibility of feed, and improving cattle productivity, especially in developing economies where low productivity means emissions levels that are many times higher than in developed economies.

These are valuable projects. Some mitigation is better than no mitigation, and some of these interventions have additional benefits that feel to me like they're even more important than the emissions benefits. For example, programs that improve cattle productivity also reduce death losses and boost incomes for pastoralists in low-income countries. That's meaningful, regardless of the climate impact.

But we also need to be clear-eyed about what's possible across the various interventions, so that we can prioritize appropriately. Even with the rosiest of assumptions, these interventions can slow but not reverse the increase in emissions from animal agriculture.

Climate analysts at Project Drawdown concluded that if every livestock-focused methane intervention worked perfectly and scaled globally, emissions would fall by about 800 million metric tons per year. That's huge: It's the equivalent of eliminating 80% of global air travel or shifting half of all cars, buses, and light trucks from gas to electric.[16]

But here's the thing: Livestock numbers are expected to rise so much by 2050 that emissions from cattle and other ruminants will increase by 1.3 billion metric tons.[17] So livestock methane interventions can cut that number to 500 million metric tons if they're universally implemented. But the overall trajectory of meat-caused emissions will still be up, not down.

Project Drawdown raises another concern: Anything close to universal implementation is extremely unlikely. About 90% of the world's four to five billion ruminants are raised entirely on pasture, not in feedlots, where feed additives and improved feed can be most easily implemented.[18] Even for feedlot cattle, incentives are lacking, and systems to ensure compliance would be costly and complex, making broad adoption for any of these interventions difficult.

But let's say it all worked perfectly for the roughly 10% of global cattle who spend part of their lives on feedlots. There's another issue: Even feedlot cattle spend the first 80% of their lives grazing, and roughly 90% of methane emissions are produced before they reach feedlots.[19] So even if a feed additive generated an impressive 80% mitigation while cows were eating it, that would net out at an actual reduction of closer to 8% overall.

The final challenge of livestock-based mitigation strategies is this: There's not a market incentive that would lead to global scaling. Even if one region or country adopts one or more of these interventions, the lack of profit motive for scaling means that every new region requires just as much effort and resources as the last. That's why the IPCC puts best-case mitigation at 200 million metric tons of mitigation per year, across all livestock-based interventions combined.[20]

200 million metric tons is a lot of emissions, and it's worth putting in the effort. But we should also be clear about what the actual mitigation potential is. It doesn't get us anywhere near that 11 billion metric tons of agricultural mitigation that WRI and the World Bank say are necessary.

One final point from Project Drawdown: Every credible climate model that meets climate targets requires that hundreds of millions of hectares of land be dedicated to nature-based carbon sequestration. A 50% shift to alternative proteins would free up roughly 650 million hectares of land. These livestock-based methane interventions would free up zero.

Two other agricultural methane mitigation strategies that generate attention are reducing food loss and waste and improving rice cultivation; according to the World Bank, the maximum mitigation potential of these strategies is 865 and 243 million metric tons, respectively.[21] That's a lot of mitigation, and there is great work happening on both interventions. Though again, that's not going to be nearly enough to reach our goals for agriculture-based mitigation. We need to use all the tools in our toolkit, including alternative proteins.

Regenerative Agriculture: Good Soil Makes Good Neighbors

If you follow NGO activity on food systems, you'll be aware of the enthusiasm for regenerative agriculture. The basic idea of regenerative farming is elegant, appealing, and right: Agriculture should work with nature instead of against it.

The Global Alliance for the Future of Food (GAFF), which is led by Anna Lappé, daughter of *Diet for a Small Planet* author Frances Moore Lappé, is leading efforts to promote regenerative farming. At COP28, GAFF released a detailed report highlighting the many benefits of regenerative practices. These include healthier soil, improved biodiversity, better yields, more resilient crops, improved nutrition and food security, and stronger rural economies.[22]

All of this is wonderful, and regenerative efforts deserve support. At GFI, we view regenerative agriculture as complementary to alternative proteins. Regenerative animal systems typically require more land, and alternative proteins free up land, which makes for a lovely symbiosis.

But once again, it's worth having some clarity about the impact on climate. While regenerative plant-based systems and regenerative systems that involve cover cropping, agroforestry, and practices that regenerate soil are all entirely positive, regenerative grazing has a few important trade-offs: Specifically, these systems require more land and cause more emissions than they mitigate or sequester.

The Food Climate Research Network, a consortium of environmental scientists from seven universities across five countries, conducted a two-year review of the evidence. Their report, *Grazed and Confused*, found that while regenerative grazing can sequester some carbon in soils, the effect is small, time-limited, reversible, and significantly outweighed by the green-house gases that livestock produce.[23]

Project Drawdown director Jonathan Foley points out that regenerative grazing in particular requires more animals grazing for more time on more land to create the same amount of meat, which swamps any sequestration potential and leads to more, rather than less, emissions.[24] In their comprehensive review of the science, WRI found that organic, grass-fed, and regenerative methods all increased land use and climate emissions relative to conventional beef systems.[25] Finally, the International Union for Conservation of Nature warns that because it tends to require more land, regenerative farming can drive deforestation.[26]

Foley still supports regenerative agriculture. He just wants to make sure there is clarity around the trade-offs: For plant production especially, regenerative practices are vastly preferable to conventional farming, and even for livestock, the improvements for farmers, ecosystems, soil health, and animals are all deeply meaningful, even if they don't reduce climate emissions.

We need an all-of-the-above strategy that takes advantage of complementary solutions. Regenerative agriculture could work well with alternative meats to address multiple problems at once. Research from the UK environmental think tank Green Alliance shows that shifting two-thirds of meat and dairy to alternative meat and dairy across 10 European countries could free up so much land that with just a third of that newly available acreage, regenerative agriculture could quadruple its footprint. More on this in chapter 10.

Life Cycle Data: Alt Meats Deliver

Spend time in climate circles, and you'll hear someone suggest it's a good idea to swap beef for chicken. It's true that chicken has a smaller environmental footprint than beef, but remember those eight plates of pasta I tossed in the trash? That was an illustration of chicken's inefficiency: 900

calories fed to a chicken to create 100 calories of chicken meat. And then add the feed mills, the chicken farms, and the gas-guzzling 18-wheelers that are trucking that feed to the farm and the animals to slaughter.

Crunch the numbers, and you find that conventional chicken from the most climate-friendly chicken farms produces roughly three times the emissions of plant-based chicken. Relative to global averages, plant-based chicken delivers less than one-tenth the climate impact of conventional chicken. Beef, pork, and farmed fish do even better.[27]

Producing this conventional meat	Instead of this alternative meat	Results in this much environmental harm relative to alt meat production	
		Climate Emissions	Land use
Conventional Chicken	Plant-based Chicken	303% (3x)	490% (4.9x)
Conventional Pork	Plant-based Pork	833% (8.3x)	250% (2.5x)
Conventional Beef	Plant-based Beef	1667% (16.7x)	1111% (11.1x)
Conventional Chicken	Cultivated Chicken	150% (1.5x)	256% (2.6x)
Conventional Pork	Cultivated Pork	173% (1.7x)	1756% (17.6x)
Conventional Beef	Cultivated Beef	995% (10x)	1756% (17.6x)

Cultivated meat is newer, so the early numbers are less dramatic but still impressive. Where renewable energy is used for both production systems, conventional chicken production produces about 50% more emissions than a small-scale cultivated meat facility; cultivated pork and beef compare even better to their conventional counterparts.[28] And again, that's relative to the lowest-emissions chicken, pig, and cattle farms—using renewable energy. As production of cultivated meat scales up, efficiency improves, and emissions numbers go down. Plus, the cultivated meat modeling used extremely conservative assumptions, four of which you can check out in the footnote at the bottom of this page.*

And that's before you consider the land use benefits, which I'll discuss more in a moment, other than to say this: Plant-based chicken requires about one-fifth of the land required for conventional chicken, and culti-vated chicken requires about one-third of the land. The land use benefits for plant-based and cultivated beef and pork are even greater. Factor in the decreased pressure on forests and the potential for land-based carbon sequestration on freed-up acreage, and the climate benefit of a shift from conventional to plant-based or cultivated chicken looks even stronger.

For comparison, switching from a gas-powered car to a plug-in hybrid cuts emissions by about a third, and going fully electric cuts emissions roughly in half—with zero land use or biodiversity benefits. So even at this early stage and even compared to the least emitting livestock systems, switching from conventional to plant-based or cultivated chicken will roughly match or ex-ceed the climate benefit of switching from a gas-powered car to an EV.

From LCAs to Global Impact

When you apply the life cycle analysis numbers to global production, the climate benefits really pop: McKinsey analysis for ClimateWorks

---

* Pelle Sinke, Elliot Swartz, Hermes Sanctorum, Coen van der Giesen, and Ingrid Odegard, "Ex-Ante Life Cycle Assessment of Commercial-Scale Cultivated Meat Production in 2030," *The International Journal of Life Cycle Assessment* 28, no. 3 (2023): 234–254, https://doi.org/10.1007/s11367-022-02128-8 (they assume no cell growth on scaffolds, even though one study found 100% cell growth during this time; they assume that media cannot be recycled, even though some com-panies are already doing it; they assume a cell doubling time of 30 hours and differentiation of 10 days—right now, companies are hitting cell doubling at 15–20 hours and differentiation in five days; and they assume significant energy needs for cooling of cultivators, even though many experts don't believe cooling will be required at all).

Foundation and the UK government estimates that at 50% global adoption of alternative proteins, we could avoid five billion metric tons of carbon emissions annually. The analysis explicitly notes that this number is conservative, since it does not include the climate benefits of sparing forests or the sequestration opportunities from 640 million hectares of freed up land.[29] Boston Consulting Group landed on a similar number.[30]

Agricultural economists at the International Institute for Applied Systems Analysis (IIASA) modeled 50% plant-based meat and milk adoption and found a mitigation potential of three billion metric tons annually, plus another 3.3 billion metric tons of mitigation potential from the 650 million hectares of land that their analysis determined would be spared from the shift. Their analysis is also conservative, since they assumed that leftover carbohydrates and oils from processing would be thrown away rather than repurposed, which is not—of course—what will actually happen. The authors note this explicitly: "An efficient scenario is more economically likely and assumes a market for co-products (as occurs in agricultural processing today)."[31]

Let's compare these numbers to airplanes and automobiles—emissions reduction and land sparing:

- Shifting 50% of animal protein production to alternatives: roughly 5–6.3 billion metric tons. Land sparing: roughly 640–650 million hectares.
- Shifting 10% of animal protein production to alternatives: roughly 1–1.25 billion metric tons. Land sparing: At least 125 million hectares.
- Eliminating air travel: roughly one billion metric tons. Land sparing: Zero.
- Shifting 100% of cars, buses, and light trucks to electric: roughly 1.5 billion metric tons. Land sparing: Zero.

All these numbers use a 100-year multiplier for methane's impact; if we use a 20-year multiplier, the mitigation potential of alt proteins relative to air travel and electric vehicles more than doubles.

Alt Meats: The "Electrify Everything" of Food

In 2020, the consultancy Rhodium Group approached GFI with an exciting request: Would GFI be willing to help them design policy and corporate engagement strategies focused on alternative proteins? The client, we learned under a nondisclosure agreement, was Bill Gates, who was preparing to launch new climate action playbooks alongside his new book, *How to Avoid a Climate Disaster.* When the book launched in 2021, the NDA lifted, and Bill gave GFI a public shoutout on LinkedIn: "We need concrete, practical solutions that give consumers choices while reducing emissions—The Good Food Institute does just that."

In the book, Gates explains his enthusiasm for plant-based and cultivated meat. For starters, alt meats are agriculture's answer to the climate mantra: "Electrify everything." As environmentalist Bill McKibben has noted, "the key to [meeting climate goals] is to electrify as much human activity as possible."[32]

As Gates puts it in the book: "Cultivated meat has all the same fat, muscles, and tendons as any animal . . . All this can be done with little or no greenhouse gas emissions, aside from the electricity you need to power the [facilities] where the process is done."

People sometimes point out that cultivated meat will require a lot of electricity, and while that may be true in the short term, it also eliminates the world's number one source of methane and slashes nitrous oxide emissions, making that an easy climate trade, as the numbers above make clear.

Think about it this way: The goal of EVs is to shift emissions footprint from carbon-spewing gasoline to electricity, which can be decarbonized. In the exact same way, alternative meats shift meat's emissions footprint from the super pollutants methane and nitrous oxide to electricity, which can be decarbonized as global electricity shifts more and more to renewables.*

Deforestation, Conservation, Land Use, and Sequestration

Another key environmental solution that is enabled by alternative proteins is nature preservation, which includes reducing deforestation. These issues are really two sides of the climate mitigation coin, since every serious model that keeps climate change within planetary boundaries requires both that deforestation be halted and that north of a billion tons of carbon be sequestered through ecosystem restoration. Without a major dip in the consumption of conventional meat, neither goal is achievable.

Think about this: Roughly half of the world's habitable land is used for agriculture, and about 80% of agricultural land—roughly 3 billion hectares—is used to either graze cattle and other ruminants or to grow feed crops for chickens, pigs, farmed fish, and other animals. Recall that according to WRI's analysis, we'll need another 3.3

---

* Oskaras Alšauskas, Elizabeth Connelly, Mathilde Huismans, Ethan Jenness, Javier Jorquera Copier, Jean-Baptiste Le Marois et al., *Global EV Outlook 2024: Moving Towards Increased Affordability* (International Energy Agency, 2024), https://iea.blob.core.windows.net/assets/a9e3544b-0b12-4e15-b407-65f5c8ce1b5f/GlobalEVOutlook2024.pdf, 150. (According to projections by the International Energy Agency, between 2023 and 2035—so over just 12 years—energy needs for EVs will be up by a factor of 20 globally and by a factor of 25 in the United States, to 10% and 16% of all global and US electricity, respectively. In both cases, this is not a difficult climate trade.)

billion hectares of land for agriculture if conventional animal agriculture continues its current trajectory and we don't increase agricultural productivity.

In this scenario, we will "clear most of the world's remaining forests" and "wipe out thousands more species."[33]

The top two drivers of deforestation globally are grazing cattle and growing soy for animal feed.[34] Many climate advocates know soy is tied to deforestation, and they know that 80% of soy is fed to farm animals; they often assume it's going mostly to cattle, and you'll often hear environmentalists say that deforestation is caused by cattle and their feed. That's incorrect. In fact, less than 1% of global soy is consumed by beef cattle. That soy is mostly fed to chickens, pigs, and farmed fish. That means when we talk about soy's environmental impact, we should be thinking about poultry, pigs, and farmed fish, not cows.

Considering the link between meat production and deforestation and considering that global meat production hits a new record every single year, it's not surprising that the world hasn't made much progress on deforestation goals. In 2021, more than 140 governments around the world pledged to reverse deforestation by 2030. Over the next few years, deforestation held steady. Then in 2024, "the tropics lost a record-shattering 6.7 million hectares of primary rainforest . . . That's more than any other year in at least the last two decades." WRI's overall assessment is not rosy: "Of the 20 countries with the largest area of primary forest, 17 have higher primary forest loss today than when the agreement was signed."[35]

Trying to halt deforestation in a world that's going to require another 3.3 billion hectares of land due to increased meat consumption doesn't feel likely. It's clear that a shift toward alt meats should be a part of our strategy to save the world's forests.

Soy as animal feed, by animal

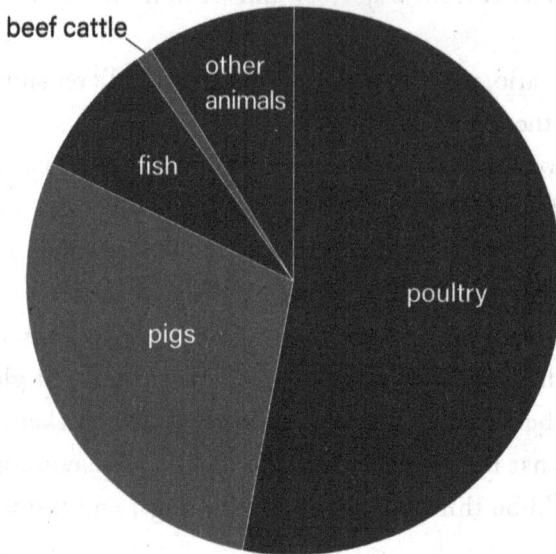

Other Environmental Harms

Recall that in 2006, when the world was consuming 104 million metric tons less meat than it does today, *Livestock's Long Shadow* found that animal agriculture is one of the top three causes of most environmental problems, at every scale from local to global. This includes not just climate change and deforestation, but also land degradation, air and water pollution, water scarcity, and biodiversity loss.[36]

In the United States, there's a steady stream of articles that underline the point: chicken farms polluting the Illinois River Watershed, leading the state of Oklahoma to sue poultry producers; a poultry slaughterhouse in Delaware contaminating drinking water and making people sick; air pollution from pig farms so bad it's visible from space. A story in the *New York Times* indicts poultry and dairy farms for draining freshwater supplies across the country, noting that: "the toll on aquifers, which supply 90% of America's water systems, has been devastating."[37]

Because their production is so much more efficient, alternative meats can help with all these environmental harms.[38]

Decimating Aquatic Life and Ecosystems

Most of the above analysis focuses on land animals, but the global seafood industry comes with its own parade of horribles. Both commercial fishing and fish farming cause significant environmental damage.

For example, about 60% of fish stocks are fished to their maximum sustainable limit, and more than one-third are overfished.[39] Bottom trawling—the method responsible for more than a quarter of all wild-caught fish—involves dragging massive nets across the ocean floor, destroying everything in their path. It's the underwater equivalent of clear-cutting a forest.[40]

The ocean also absorbs atmospheric carbon that would otherwise heat the planet even faster. Scientists at Imperial College London and the UK's Natural Environment Research Council warn that commercial fishing has profoundly altered marine ecosystems, reducing the ocean's ability to play this critical role. And bottom trawling releases an estimated 370 million metric tons of carbon annually—about double the emissions from US air travel.[41]

The aquaculture industry often frames itself as a more environmentally friendly way to supply the world's growing demand for seafood, but aquaculture is a top contributor to deforestation and ocean pollution. Like land-based animal farming, aquaculture is incredibly inefficient, requiring 10 calories of feed to produce one calorie of fish.[42] The number one source of those calories is soy, which is one of the world's top contributors to deforestation.[43] Fish farming is expected to double through 2050, and as it does, so will its soy demand and its adverse impact on rainforests.[44]

Meanwhile, the fishing industry generates staggering amounts of ecosystem-destroying pollution in the form of "ghost gear": abandoned nets, lines, and traps left to drift in the ocean. Ghost gear is the deadliest form of marine plastic according to the World Wildlife Foundation, which found that discarded fishing gear harms two-thirds of marine mammals, half of seabirds, and all species of sea turtles.[45] I don't know about you, but to me, that feels like a truly unfathomable adverse impact on wild animals.

When researchers examined entangled marine animals along the California coast, they found that more than 90% of entanglements were caused by fishing gear, in some cases sending animals to the brink of extinction.[46] This devastation is particularly evident in the subtropical oceanic gyre in the North Pacific Ocean (aka, the North Pacific Garbage Patch): Tens of thousands of metric tons of floating plastic, spread over an area three times the size of France. Three-quarters of that plastic is discarded fishing gear.[47]

A shift toward plant-based and cultivated seafood could eliminate all of this.

———

We're at code red for the climate. We're losing forests faster than ever. Neither crisis can be solved if conventional meat production continues on its current trajectory. WRI says we're rushing headlong toward a world with no forests.

For decades, agriculture has been a third rail in policy discussions related to climate and nature, because the solutions all seemed either too small or too intractable. As Bill Gates said to Mark Rober in this chapter's introduction, Bill thought that meat production would be the toughest of all climate challenges.

What we need, Gates realized, is an agriculture-based solution that could scale.

This is the critical point where alt proteins really shine. McKinsey economists working for the ClimateWorks Foundation and Global Methane Hub analyzed the most common agriculture-based methane mitigation interventions: everything from lower-methane rice farming to slashing food loss and waste to the range of interventions focused on livestock.

The only intervention that played the markets card effectively was alternative proteins, which provided 98% of the $700 billion in economic benefits across all the analyzed interventions.[48] What that means is that once we have products at price and taste parity, markets will be incentivized to send them up the S-curve of adoption.*

We need an all-of-the-above plan for reaching climate and nature goals. All climate mitigation is good mitigation. All forest preservation is good forest preservation.

But without alternative proteins, the science is very, very clear: Climate and deforestation efforts will fall short of our goals, just like energy transition goals can't succeed without renewables.

Alt meats can slash emissions, preserve forests, and free up land for nature-based sequestration. It's an essential tool in our climate mitigation and nature preservation toolkits.

---

* To be clear, I'm not saying this will happen automatically; I'm saying that at price and taste parity, market incentives kick in, which will almost certainly focus on accentuating the health and safety benefits of alt meats (see chapters 6 and 7) and allaying any concerns with regard to processing, naturalness, etc.

CHAPTER THREE

Saving Modern Medicine

A post-antibiotic era means, in effect, an end to modern medicine as we know it. Things as common as strep throat or a child's scratched knee could once again kill. Some sophisticated interventions, like hip replacements, organ transplants, cancer chemotherapy, and care of preterm infants, would become far more difficult or even too dangerous to undertake.

— Dr. Margaret Chan, Director-General,
World Health Organization (2012)[1]

In the century since Alexander Fleming stumbled across penicillin in a laboratory in London, antibiotics have become a mainstay of medicine, transforming once-deadly infections into treatable and curable conditions. Antimicrobial resistance threatens to unwind that progress, making it without question one of the most pressing health challenges of our time.

— Dr. Tedros Adhanom Ghebreyesus, Director-General,
World Health Organization (2024)[2]

The Definition of Insanity?

The comedian Robin Williams once joked about policing differences between the United States and the UK. In the United States, he said, if you run from the police, they shoot. In the UK, the police don't carry guns, so if you run, they yell, "Stop!" And if you don't stop? Then things get serious: They yell "Stop!" again.

I often think about that joke when I consider the past 25 years of public health warnings about antimicrobial resistance (AMR). For decades, scientists have been shouting "Stop!" about the misuse of antibiotics in farm animals. Industry doesn't stop. So the health community, powerless to enforce real change, just yells "Stop!" again. And again. And again.

Each year, the global pharmaceutical industry produces about 136,000 metric tons of antibiotics. These drugs are critical to human health. Yet less than a third of them are used to treat sick people. More than 70%—nearly 100,000 metric tons—are fed to farm animals, primarily to speed up growth or to prevent disease in stressful, overcrowded conditions.[3]

Here's how AMR works: Every time bacteria are exposed to antibiotics, a few survive by mutating. Over time, these survivors multiply, becoming stronger, more resistant, and eventually immune to treatment. It's like playing a video game where the enemy adapts to your every move until your most powerful weapons are useless. This can happen in humans; it can also happen in farm animals.

Already, many bacteria have mutated beyond the capacity of existing drugs to kill them. These bacteria are called "superbugs."

The *Lancet's* 2022 comprehensive review found that antimicrobial-resistant infections killed about 1.27 million people in 2019—more than HIV/AIDS or malaria—and contributed to another 3.5 million deaths. By 2050, experts predict that AMR will contribute to more than eight million deaths worldwide, killing more than two million people per year outright.[4][5]

In the United States, the CDC reports more than 2.8 million super-bug infections each year and more than 35,000 deaths.[6] Even nonfatal infections can lead to prolonged illness, disability, and higher risk of complications from surgeries and other medical procedures.

Antibiotics are used to treat everything from minor tooth infections to life-threatening conditions like pneumonia and meningitis. They are also a critical element of post-surgical care—protecting patients from infection after stitches from sports injuries, knee or hip replacements, and organ transplants. They fight off urinary tract infections and protect cancer patients undergoing chemotherapy and preterm infants in neonatal units.

In pre-antibiotic times, simple cuts and scrapes could lead to infection that would require amputation. If you were poor and couldn't afford a doctor, you would likely die. Even if you weren't poor, death from infection was incredibly common.

In 2012, WHO Director-General Dr. Margaret Chan delivered a keynote speech at an international conference on AMR:

"A post-antibiotic era means, in effect, an end to modern medicine as we know it. Things as common as strep throat or a child's scratched knee could once again kill. Sophisticated interventions like hip replacements, organ transplants, cancer chemotherapy, and care of preterm infants would become far more difficult or even too dangerous to undertake." The threat, she said, was "global, extremely serious, and growing."

One example of what could be coming is a huge increase in maternal mortality: A 2014 report for the UK government explained that "the 20th century saw childbirth in high income countries move from being something that carried significant risk to something that we take for granted as being safe: The world witnessed a 50-fold decrease in maternal deaths over the course of that century. Much of this progress could risk being undermined if AMR is allowed to continue rising significantly."

What antibiotics have given us, AMR threatens to take away.

I am prone to bike accidents and tooth infections: In the past decade, I've broken my collar bone, impaled my heel on a car license plate down to the bone, and smashed up my face so badly that I required 13 stitches. Liz Specht, GFI's former VP of science and technology, jokes that GFI needed a clear succession plan, since there was always a reasonable chance that I'd either die or become incapacitated in a bike accident. As I type this, it occurs to me that in a world without antibiotics, I might already be dead.

I'll spare you my tooth infection stories other than to say that I'm familiar with dental offices all over the world and travel with pain medications and amoxicillin, which turned me around on western medicine. For the first 40 years of my life, I really only went to the doctor when I was physically injured, and even then, I did my best to avoid prescription medication. Not anymore.

Factory Farms: Superbug Factories

The use of medically important antibiotics in farm animals is the most avoidable driver of AMR: As noted, more than 70% of antibiotics critical for human medicine are used not to treat people but to promote faster growth or prevent disease in animals raised for food. In some countries, especially in the developing world, the figure is even higher—up to 80%.

Why so much?

Simple: Farm animals are given antibiotics throughout their lives to make them grow more quickly or to keep them alive in squalid conditions, while humans take them only for a few days at a time, to treat some specific condition. Once that reality sinks in, 70% feels low. Since reporting is spotty, we can't know for sure, but researchers have consistently found that actual use in farm animals is much higher than reported.[7]

And antibiotic use in farm animals isn't just widespread—it's rising.[8]

Aquaculture's Antibiotic Addiction

It's not just land animals: Aquaculture, or fish farming, accounts for an additional 10,000 metric tons of antibiotics each year, a figure projected to rise to 13,600 metric tons by 2030.[9] Fish farms, often crowded and poorly regulated, require vast quantities of antimicrobials to keep infections at bay.

A 2023 UN Environment Programme (UNEP) report warned that aquaculture creates ideal conditions for the creation of superbugs. Fish farms house a far greater variety of species than land-based farms, allowing bacteria to more easily jump between hosts, exchange genetic material, and develop resistance. These "genetic hotspots" also accelerate the evolution of dangerous pathogens, which is the subject of the next chapter.[10]

There's vanishingly little meaningful oversight of drug use on fish farms. About 90% of global aquaculture takes place in China and other developing nations, where regulations are weak and enforcement even weaker. Similarly, more than 90% of seafood sold in the United States is imported—mostly from Asia—under standards that most American consumers would find unacceptable.[11]

Antibiotic pollution compounds the risk: In human medicine, antibiotics flushed from our bodies typically pass through wastewater treatment plants. In agriculture, about 75% of farm animal antibiotics are excreted into the environment untreated.[12] In aquaculture, it's even worse, since much aquaculture happens not inland like you might be picturing but in lakes and oceans where the fish are contained by netting. About 80% of antibiotics for aquaculture disseminate untreated into nearby ecosystems.[13]

The result? Increased parasites and pathogens in our lakes, streams, rivers, and oceans. For the animals who call these aquatic environments home, the effects can be severe: birth and developmental defects, sicknesses, and death. The drugs also end up in fish and then in the bodies of human beings, with human health impacts that include reproductive problems, muscle weakness, and an increased risk for inflammatory disorders.[14]

The Public Health Community: Yelling Stop, Decade After Decade

In September 2024, the head of the World Health Organization, Dr. Tedros Adhanom Ghebreyesus, called AMR "without question one of the most pressing health challenges of our time." He called for the global health community to come together and slash unnecessary human prescriptions, slash use in animals, improve tracking and education, and fund research into new antibiotics.

His warning was covered worldwide: Global health authorities were getting serious about AMR.

Um, no.

WHO's press release did not mention that the public health community had been making similar statements for decades, without much in the way of meaningful progress. Neither did coverage in the popular press.

Antibiotic treatment in humans started in the 1940s, and by the early 1950s, the drugs were being used to make farm animals grow faster. Farmers quickly learned that drugged animals not only grew more quickly, but they also died in far fewer numbers. Plus, they could withstand almost unfathomably crowded conditions that would, absent the drugs, have led to skyrocketing death rates.

"Hey, look how many drugged up pigs or chickens you can cram into a filthy barn! Thousands of pigs! Tens of thousands of chickens!" Industrial animal farming was born.

If you're looking for the full history of how antibiotics fueled the rise of both industrial farming and superbugs, I recommend Maryn McKenna's brilliant book *Big Chicken: The Incredible Story of How Antibiotics Created Modern Agriculture and Changed the Way the World Eats.* It's a sobering deep dive into how drugging animals has sent a public health crisis into overdrive.

All the way back in 1953, during a hearing in the British Parliament, Conservative MP Alan Gomme-Duncan called it madness to feed drugs to pigs, arguing that the animals should be treated better rather than drugged prophylactically. He voiced a simple and rhetorical question: "Why give penicillin to pigs to fatten them? Why not give them good food, as God [intended]?"[15]

That same decade, British microbiologist Ephraim Anderson warned of the human risks associated with using antibiotics in livestock. His findings appeared in *The Lancet, The British Medical Journal,* and *Nature*—some of the most prestigious scientific publications in the world. The media picked up his warnings, but nothing changed.

Essentially, Anderson yelled "stop!" as loudly as he could. And then—nothing happened.

Fast forward to 1998. That year, the World Health Assembly—the decision-making body of the World Health Organization—adopted a resolution acknowledging antimicrobial resistance as a global health threat and calling for a concerted global effort to address the issue.[16]

Their report identified two main causes:

1. Overuse of antibiotics in animal agriculture.
2. Overprescription of antibiotics in human medicine.

And recommended four key actions:

1. Surveillance to track resistance trends.
2. Education and regulation to reduce unnecessary use in humans.
3. Reduced use in farm animals.
4. Research to develop new treatments.

Sound familiar? It should; it's been the global health community's prescription ever since, including a few paragraphs up, from the head of the WHO in 2024. For 30 years, the world's public health authorities have produced a steady stream of reports, conferences, and policy packages urging the world to act—essentially issuing the same basic analysis again and again and again. Meanwhile, antibiotic use in both humans and farm animals has soared. Superbugs have flourished. And new drug development has stalled.

A few notable moments along the way:

- **2000**: WHO urged countries to ban antibiotics for farm animal growth promotion.[17]
- **2001**: WHO warned that the antibiotic pipeline was running dry and called for urgent action, including decreased use in farm animals, "to avert future disaster."[18]
- **2007**: Agricultural researchers from the United States and Sweden called for a "phasing-out of the use of antimicrobial growth promotants for livestock and fish production."[19]
- **2011**: Doctors from around the world published a plea in *The Lancet* for an end to antibiotics use in livestock and fish farms. "There is no longer time for silence and complacency."[20]
- **2016**: AMR became the fourth health topic discussed at the UN General Assembly. Dr. Chan declared that "we are running out of time." FAO Director General Dr José Graziano declared: "Agriculture must shoulder its share of responsibility, both by using antimicrobials more responsibly and by cutting down on the need to use them." WHO titled its press release: "At UN, global leaders commit to act on antimicrobial resistance."[21] Spoiler: Nothing changed.
- **2017**: WHO again called for an end to routine antibiotic use in healthy animals, based on "decades of expert reports and

evaluations of the role of agricultural antibiotic use in the increasing threat of antibiotic resistance."[22]

- **2024:** A group of 11 scientists from eight countries declared in *The Lancet* that AMR jeopardizes global targets around "child survival, healthy aging, poverty reduction, and food security." Their solution: a 20% reduction in human use and a 30% reduction in animal use.[23]
- **2024:** The Global Leaders Group on AMR warned of a looming human and economic catastrophe without immediate, bold and urgent action, including the elimination of routine use of antibiotics in agriculture.[24]

I'm not sure what I expect the medical community to do, since it doesn't have the power to force compliance with any of its recommendations. I suppose the main thing I'd like to see is a lot less self-congratulation. In 2024, the WHO press release declared, "World leaders commit to decisive action on antimicrobial resistance."

That's pretty close to the 2000 claim from WHO that it had a "comprehensive global strategy for the containment of antimicrobial resistance." And in 2016: "At UN, global leaders commit to act on antimicrobial resistance."

If you're just going to keep yelling "stop!" and if that yelling hasn't accomplished much, at the very least, you might want to stop sending out press releases claiming that the problem is under control.

———

Despite all the talking, meeting, and declaring, antibiotic use in farm animals continues to rise. Gains made mostly in Europe and to a lesser extent in the United States have been overwhelmed by skyrocketing antibiotic use elsewhere. Researchers estimate that by 2030, Brazil, Russia,

India, China, and South Africa will double their farm animal antibiotic use compared to 2010 levels. And in China—the world's largest consumer of veterinary antibiotics—actual use may already be almost double official government figures.[25]

But superbugs don't respect national boundaries. A superbug that develops anywhere can spread across the globe and infect people everywhere.

How Real is the Progress in Europe and the United States?

When I shared this chapter with my friends at Farm Forward, a nonprofit organization that monitors the impact of industrial farming on global health and advocates for change, longtime executive director Andrew deCoriolis pointed out that even the supposed progress in Europe and the United States may not be progress at all. In most cases, decreased use of medically relevant antibiotics is accomplished through a ramping up of other drugs.

Sure enough, when I reread the journal article about global trends in antibiotics use, I found that in addition to the prediction of a sharp increase in the use of medically relevant antibiotics globally, what limited progress has been achieved has mostly involved a shift "from medically important antibiotics to the use of ionophores which are not included in [antimicrobial use] data published by most countries." The UK, for example, cut its annual use of medically relevant antibiotics by 80 metric tons between 2013 and 2017, even as it increased use of ionophores by 72 metric tons.

Andrew noted that these drugs are even less well-monitored globally and that their effects are poorly understood. While we don't know if they contribute to AMR, they certainly contribute to antibiotic pollution. Andrew explained to me that farmers' hands are tied, especially in developing economies.

The basic problem is that modern farm animal genetics and modern intensive farming add up to massive animal death losses if drugs are withdrawn. For example, the US poultry industry slaughters more than nine billion birds per year, and almost all of them are intensively bred in ways that compromise the birds' immunity. Drugs become essential. "When companies try to transition away from raising animals with the help of pharmaceuticals," Andrew shared, "what they mostly discover is that it's unprofitable."

Here again, regenerative farming practices could help, since well-treated animals with robust immune systems don't need drugs to survive their daily existence like animals in industrial systems. Farm Forward advocates a shift that's a few steps better than standard regenerative agriculture: Heritage breed animals. These breeds have much greater genetic diversity, and they are raised in hygienic conditions. That said, meat from heritage breeds costs four times as much as industrial animal meat, if you can find it at all.

When the Drugs Stop Working

What about the other consistent recommendations from the public health community over the past 30 years: decreased use of antibiotics in humans and funding the development of new antibiotics? Sadly, these interventions are not working either.

Human antibiotic use has been up consistently for decades, and that shows no signs of slowing down. Per capita antibiotic use in human medicine increased by 65% between 2000 and 2015, by 16% between 2016 and 2023, and is expected to be up about 50% between 2023 and 2030.[26]

But that is far from the whole story: In many ways, antibiotic use is an equity issue, with rich countries vastly overprescribing these critical drugs, even as lack of access is a significant cause of death in many low-income countries. Indeed, per capita use in some developed countries is nine times higher than in some developing countries.[27]

We're also not making much progress on the development of new antibiotics. Drug development is expensive, and drug approvals for human medicine can take a decade or more. Combine that with the very high failure rate of antibiotic drug research and the very short courses of antibiotics when prescribed, and the private sector incentives simply don't line up.[28] In 2023, the World Health Organization issued a grim progress report: The global pipeline of new antibiotics remained "insufficient," most major pharmaceutical companies had abandoned antibiotic research altogether, and the few small biotech firms still working in the space struggled to survive financially.[29] [30]

Governments in G7 countries devote about $1.4 billion annually to antimicrobial R&D. But experts agree it's not enough. The Global Leadership Group on AMR highlighted a shrinking pool of scientists working on antibiotic discovery, a lack of profitable markets for new antibiotics, and a wave of bankruptcies among small firms.[31]

In short: We are running out of weapons—and out of time.

Alt Meats: Antibiotic-Free, by Design

The global health community has spent 70 years yelling "Stop!" over and over in response to the use of human-relevant drugs in farm animals.

This has worked in some European countries and to a lesser degree in the United States, though if the UK is any indication, what may be happening is more a shift from medically relevant drugs that we know cause extensive harm to non-medically relevant drugs that have been subjected to much less research.

Globally, even this intervention faces serious barriers: In many regions, drug-fueled growth promotion boosts farmer profits. For farmers in developing economies, giving up antibiotics (or even shifting which antibiotics are used) without compensation is unrealistic. Improving sanitary conditions would require expensive changes many farmers can't afford.

And governments in many countries lack the political will—or the resources—to mandate better practices.

By contrast, alternative proteins offer a fundamentally different solution. Plants are not animals, so antibiotics are not used in plant-based meat
production. Cultivated meat production could use antibiotics, but the first
seven approved products did not.[32] And all cultivated meat trade groups
have all pledged to keep cultivated meat antibiotic-free. For cultivated
meat, this aligns to profitability: Antibiotics inhibit growth in cultivators.

As global meat consumption rises, the choice is stark: We can continue relying on drug-intensive industrial farming, accelerating the march
toward a post-antibiotic era. Or we can shift toward alternative proteins
that dramatically reduce AMR risks while satisfying global demand for
meat.

The future of modern medicine may depend on the path we choose.

CHAPTER FOUR

Preventing the Next Pandemic

Animal agriculture expansion and industrialization [are] linked to increased risk of zoonoses emergence. The majority (70%) of emerging infectious diseases and almost all known pandemics (e.g., influenza, HIV/AIDS, COVID-19) are zoonoses.

— United Nations Environment Programme, 2023[1]

The Pandemic Powder Keg

In 2007, I read *Bird Flu* by Dr. Michael Greger, and I almost couldn't believe what I was reading. Over more than 400 deeply researched pages, Dr. Greger laid out a chilling case: Most contagious diseases originate from domesticated animals, and industrial animal farming is a pandemic powder keg, waiting to explode.

Expert after expert, journal article after journal article, supported his core argument:

- Another pandemic is inevitable.
- It could be much deadlier and more transmissible than anything humanity has seen so far.
- It will most likely come from an industrial chicken or pig farm.

Dr. Greger worked closely with leading scientists to assemble the book, and one of the world's leading virologists, Kennedy Shortridge from the University of Hong Kong, wrote the foreword. Shortridge described industrial poultry farming as "an influenza accident waiting to happen" and warned that current practices "threaten to trigger a human catastrophe."

The book received positive coverage in *The Journal of the American Medical Association*, *Nature*, and other top scientific journals. It also earned glowing endorsements from dozens of public health experts, including Interior Secretary Dirk Kempthorne and CDC director Dr. Julie Gerberding.* But it didn't make much impact beyond academic and scientific circles.

After reading Dr. Greger's book, the 1918 bird flu became my go-to example of our very human tendency to forget the past. Bird flu appears to have originated not in a distant forest, but in Haskell County, Kansas. It jumped the species barrier and killed 50 to 100 million people in less than two years.[2]

For comparison: Across the globe, as many people died from the 1918 bird flu as died in either World War I or World War II.[3] And *American* death losses from the flu were more than American death losses from both world wars combined.[4]

To me, that feels quite remarkable, and if you're like me, you had no idea. Where's the Ken Burns documentary about the 1918 bird flu pandemic? By

---

* For example: leading pandemic specialists Graeme Laver from Australia, Earl Brown from Canada, and Kennedy Shortridge from the University of Hong Kong. Dr. Greger also had extensive assistance from multiple scientists with the Beijing Bureau of Agriculture and China Agricultural University.

the time I was studying agricultural economics just a few generations later, I don't think the 1918 pandemic came up. And in coverage of bird flu outbreaks ever since, pandemic risks have been downplayed or ignored.

The Two Most Likely Causes of the Next Pandemic: Increased Meat Consumption and Industrial Meat Production

When Covid-19 hit, global health experts sprang into action to figure out how to prevent the next pandemic. A team from the International Livestock Research Institute (ILRI), UNEP, and CGIAR pulled together a dozen of the world's leading zoonotic disease specialists as coauthors and another 60-plus experts across top global agencies as reviewers.* Led by ILRI's Delia Grace Randolph, they released their findings in July 2020 in a report titled *Preventing the Next Pandemic: Zoonotic Diseases and How to Break the Chain of Transmission.*

The number one risk factor for the next global pandemic? The world's growing appetite for animal meat. More meat means more animals. And more animals mean more viral risk. Many people assume pandemics are primarily the result of human interaction with wild animals. But the science says otherwise. In the report's introduction, UNEP Executive Director Inger Andersen explains that three-quarters of emerging infectious diseases—e.g., Ebola, SARS, Zika—started in animals. And she calls for "radically transforming food systems," because most zoonoses don't emerge through direct human-wildlife contact. Instead, they move through the food system.

The number-two risk: industrial-scale animal farming, for at least three reasons.

---

* For example, the World Health Organization, the UN Development Programme, the World Organization for Animal Health, and major universities.

First, the animals we raise for food are so genetically similar to one another. For example, in the United States and around the world, virtually all meat chickens come from just a few strains. The UNEP report explains why this matters:

"Genetically homogenous host populations are more vulnerable to infection than genetically diverse populations, because the latter are more likely to include individuals that better resist disease." In other words, when all animals are alike, a virus that infects one can easily infect them all.

Second, modern industrial farms are perfect breeding grounds for pathogens: Tens of thousands of chickens crammed into sheds, living in their own waste and breathing ammonia-thick air. Pigs raised in intensive confinement, their artificially massive bodies pushing their immune systems to the limit. These conditions depress the animals' immunity and allow viruses and bacteria to spread rapidly and mutate dangerously.

The third reason has to do with transport to slaughter. At the end of their short lives—roughly six weeks for chickens and six months for pigs—the animals are crammed onto trucks and shipped to slaughter, often through extreme heat in the summer or extreme cold in the winter. Animals routinely arrive at slaughter crippled, frozen, sick, and dead.[5] Chickens fall off trucks and are eaten by scavengers.

All of this increases risk. The transport process puts them into even closer contact with one another, and the physiological strain of extreme weather decreases their immunity. Finally, shipping them all over the country increases their likelihood of encountering wild animals; that allows the virus to mutate as it crosses into other species and allows those animals to spread the virus even further.

As Johns Hopkins Bloomberg School of Public Health professor of microbiology and immunology Andrew Pekosz explained to *Environmental Health Perspectives*: "This is all a numbers game. The more variants you're exposed to, the more likely it is that you'll be exposed to one with altered properties that allows for infection of a new host."[6]

Playing Russian Roulette with Humanity

Bird flu outbreaks have been happening for decades. But I never paid much attention. If you're like my friends and relatives, neither did you.

Consider the outbreak in 2015: It cost the federal government $879 million and the meat industry $3.3 billion. More than 50 million birds were culled across 15 states.[7] [8] And yet unless you were working in agriculture, you probably don't remember it. I was paying attention at the time because of GFI's early work, and yet I barely registered it. The USDA reassured the public that human risk was low, and coverage outside the agriculture-focused media was minimal.

Flash back six more years to March 2009, when swine flu emerged on a pig farm in Mexico. It jumped the species barrier and spread globally.[9] Within weeks, the WHO declared a public health emergency. The CDC released more than 10 million doses of antivirals and tens of millions of masks and respirators. And WHO declared a full pandemic by June.

By the end, the pandemic had spread to 209 countries, causing more than 60 million infections, 250,000 hospitalizations, and 10,000 deaths in the United States alone.[10] Unlike seasonal flu, 80% of deaths were among people under 65.

That said, a full-fledged pandemic that killed only 10,000 Americans? That sure feels like we dodged a bullet. *New York Times* reporter Donald McNeil agrees: "The virus and the vaccine cooperated. While the former proved highly transmissible in children, it was rarely lethal, remained susceptible to drugs, and did not mutate into an unpredictable monster."[11]

You probably don't remember that one either—I didn't, and I was teaching in inner city Baltimore at the time through Teach For America.

Schools are a notorious incubator for colds and flu, so I would have thought this kind of thing would have been noticed. It wasn't.

The reason: It's just too common. Chicken and pig flu are now endemic across global agriculture. For example, a 2020 European study sampled 2,500 pig farms and found influenza viruses on more than 50% of them. Scientists aren't shy about the implications. In their peer-reviewed *Cell Host & Microbe* paper, 21 researchers from Germany, the UK, Finland, and the United States warned: "This represents a constant threat to public health because [influenza strains] could cross the species barrier into humans, causing severe morbidity and contributing to global pandemics."[12]

The scientific community keeps sounding the alarm.

The rest of the world keeps hitting snooze.

It's worth remembering: Covid-19, devastating though it was, wasn't especially lethal. If the next pandemic mirrors 1918's virulence instead of 2019's, the consequences could be catastrophic.

And the next pandemic could happen at any time. The year 2020 didn't just bring Covid-19. It also marked the beginning of an unprecedented global spread of a highly pathogenic strain of bird flu, H5N1, which decimated wild bird populations before moving on to infect many other species across five continents.[13]

By 2023, H5N1 had infected "badgers, black bears, bobcats, coyotes, ferrets, fisher cats, foxes, leopards, opossums, pigs, raccoons, skunks, sea lions, and wild otters."[14] It spread across more than 80 countries, reaching both the Arctic and Antarctic. It infected domestic dogs. It infected bald eagles and condors. It infected polar bears.

In Europe, authorities culled mink and foxes on fur farms en masse, desperate to contain outbreaks. In the United States, tens of millions of chickens were slaughtered across 47 states, at a cost to taxpayers of more than a billion dollars.[15] And in 2024, horrifying new reports emerged:

- Mass deaths among sea lions.
- Seabirds so weakened and disoriented they could no longer fly.

- Elephant seals suffering neurological tremors.
- Cats going blind, walking in circles, and then dying.[16]

Meanwhile, public messaging stayed eerily calm. Even as H5N1 tore through mammal populations and the virus demonstrated clear capacity to jump species, the refrain remained the same: The risk to the general public was said to be low.[17]

In early 2024, a major shift occurred, when dairy workers in the United States tested positive for H5N1 infections acquired from cattle.[18] This marked a critical escalation. Mammal-to-human transmission dramatically increases the likelihood that the virus could mutate into a form capable of efficient human-to-human spread.

And the more the virus circulates among mammals, the more evolutionary chances it gets.

As *New York Times* science reporter Apoorva Mandavilli wrote in June 2024: The fact that H5N1 is not currently killing large numbers of humans "does not guarantee that [it] will remain benign if it begins to spread among people. Accumulating evidence from the animal world and data from other parts of the globe . . . suggest the opposite . . . The worry now is that as H5N1 continues to infect mammals and evolve, it may pick up the mutations needed to spread efficiently among people, setting off another pandemic."[19]

Anice Lowen, a virologist at Emory University, described the nightmare scenario:

If a person were to become simultaneously infected with H5N1 and a seasonal flu strain, the viruses could swap genetic material, giving H5N1 the ability to spread through human populations with the ease of the seasonal flu.[20]

We are playing Russian roulette here, essentially begging for a recurrence of the 1918 flu pandemic, but in a world that is at far greater risk than we have ever been in human history. Take those top two risk factors: meat consumption and industrial farming. In 1918, the world was eating about a tenth of the meat we're consuming now, and it came from a wide variety of robust genetic strains, none from genetically similar drugged animals with weak immune systems on industrial farms.*

The Experts Weigh In: Band-Aids for Bullet Wounds

By late 2024, some experts were starting to sound more urgent alarms about H5N1.

In November and December, *The New York Times* published two strong guest essays on pandemic prevention—one by David Kessler, who had led the FDA under Presidents George H. W. Bush and Bill Clinton and later served as chief science officer under Presidents Trump and Biden, and another by Princeton professor Zeynep Tufekci.[21]

In his piece, "I Ran Operation Warp Speed. I'm Concerned About Bird Flu," Dr. Kessler described the urgency of the situation:

- H5N1 had infected more than 60 dairy herds in 15 US states.
- Tens of millions of birds had been culled across 49 states.
- 55 human infections had been confirmed, mostly among farmworkers.
- Half of California's dairy farms might already harbor the virus.

---

* We only have meat consumption statistics going back to 1961, when the world was eating 71 million metric tons of meat per year—across roughly 3 billion people. Hannah Ritchie, Pablo Rosado, and Max Roser, "Meat and Dairy Production," Our World in Data, last updated December 2023, https://ourworldindata.org/meat-production. Assuming the same per capita consumption, this would indicate that 42.6 MMT of meat were consumed in 1918. This is conservative, since we know that meat consumption rises with income, and poverty rates globally fell from 59.6% to 50.24% during this time. Max Roser, "The Short History of Global Living Conditions and Why It Matters That We Know It," Our World in Data, last updated February 2024, https://ourworldindata .org/a-history-of-global-living-conditions. In the three decades after 1961, meat consumption rose by 31% to 33% per decade: 70.57 MMT in 1961; 103.59 in 1971; 137.77 in 1981; 181.05 in 1991.

Kessler closed his piece with appropriate alarm: "If the virus begins to transmit efficiently among humans, it will be very difficult to contain," and the likelihood of a pandemic will be high.

But his recommendations amounted to bandaging a bullet wound. They focused on crisis response, without addressing the systemic drivers putting us at risk in the first place:

- Pasteurize milk.
- Test milk for viral contamination.
- Accelerate research into antiviral treatments and vaccines.

How do we actually prevent a future pandemic? He doesn't say.

A few days later, the *Times* ran another guest essay, this time by Zeynep Tufekci with a title that seemed hopeful: "How to Avoid a Fore-seeable Catastrophe." It ran in the Sunday edition, and the front page of the opinion section included this hopeful blurb: "The best gift President Biden could leave the country and the world? Action on bird flu."

The piece was good and important, but it didn't live up to the headline. "Almost five years after Covid blew into our lives," Tufekci began, "the main thing standing between us and the next global pandemic is luck. And with the advent of flu season, that luck may well be running out."

Okay, so that's quite the opening! Unfortunately, Tufekci's suggestions also stayed exclusively in the containment zone:

- Mandatory testing of cows, milk, and farmworkers.
- Isolation of infected cattle herds.
- Speeding up vaccine development—for cows and for humans.

Good ideas, but again: no mention of reducing animal density, shifting toward a greater variety of robust animal breeds, or fundamentally rethinking how we produce meat.

Even as H5N1 spread across continents, infected a growing number of mammals, and breached the human species barrier, the leading public discussions remained locked on mitigating the impact of a pandemic, not preventing it.

The root causes—the growing demand for animal protein, the spread of industrial farming, the collapse of biosecurity safeguards—went almost entirely unmentioned.

The bullets are about to fly; the experts are suggesting that we stock up on Band-Aids.

Racing Toward the Next Pandemic

It would be comforting to believe that if global leaders just understood pandemic risk more clearly, they would act.

But the scientific community has already been clear. Crystal clear.

In 2020, the *Preventing the Next Pandemic* report from CGIAR, ILRI, and UNEP laid out seven primary risk factors that increase the chance of another zoonotic pandemic:

1. Increased demand for animal protein (from farm animals).
2. Unsustainable intensification of animal agriculture.
3. Unsustainable exploitation of natural resources.
4. Increased travel and transportation.
5. Expanding and increasingly complex food supply chains.
6. Exploitation of wildlife.
7. Climate change.

This wasn't new information. Indeed, WHO and FAO released a report in 2004 that listed "increased demand for animal protein" as the top risk factor for the emergence and spread of zoonotic diseases.[22] And the

American Public Health Association called for a moratorium on industrial animal farms in 2003 and reiterated that call in 2019.[23]

The world is not taking the actions required to mitigate risk, and the trends across all seven drivers are getting worse, not better.

- Demand for meat from farm animals continues to climb globally, with projections showing a new global record every year through 2050.
- Industrialized farming is expanding into new regions at an unprecedented pace, especially across Asia, Africa, and Latin America.
- Natural ecosystems are being cleared to create grazing land and to grow animal feed, accelerating deforestation and biodiversity loss.
- Global travel and trade—by people, animals, and goods—continue to intensify, stretching biosecurity safeguards to the breaking point.
- Food supply chains are becoming longer, more diffuse, and more fragile, increasing opportunities for pathogen spillover and rapid spread.
- Wildlife exploitation through hunting, trafficking, and habitat encroachment persists despite international agreements and NGO efforts.
- Climate change compounds every one of these threats, driving migration, stressing ecosystems, and reshuffling disease vectors around the world.

Inger Andersen, Executive Director of UNEP, explained in the report's foreword that "governments, citizens, and the private sector must work together" to shift these trends. Jimmy Smith, Director General of ILRI, stressed that until now, most efforts to control zoonotic diseases have been reactive rather than preventive. Covid-19, both argued, is a wake-up call, a reminder that patching up symptoms is no substitute for solving root causes.

Sadly, the trends are not shifting; they're accelerating. Experts continue to recommend reactive rather than preventative measures in response. And the whole world—even the experts—did not heed the wake-up call. Everyone appears to have gone back to sleep.

When I spoke with Dr. Greger in early 2025, he told me he had hoped that Covid-19 would revive interest in his book on bird flu. It didn't. His *How Not to Die* books are bestsellers. A reissue of *Bird Flu* was not. It seems the world simply doesn't want to think too hard about the pandemic powder keg we live atop because of how we make meat.

But regardless of whether we want to think about it or not, the world's pandemic experts are clear: Another pandemic is a matter of when, not if. We could be looking at something more like 1918 than 2019 in terms of both transmissibility and lethality; if so, it would kill hundreds of millions of people and sicken far more.

And the most likely cause is an industrial chicken or pig farm.

Back to the powder keg metaphor from this chapter's opening: Imagine a massive powder keg, sitting in the middle of a small city. It's old, it's volatile, it's increasingly unstable. Experts warn that a single spark—a tossed cigarette, a lightning strike, a stray bottle rocket—could set off an explosion that levels the entire town.

Most people ignore the powder keg, but some experts are concerned. They urge action: Get buckets of water ready. Train a fire brigade. Monitor the skies for potential lightning storms. Make sure hospitals are ready with doctors and nurses who are trained to treat burn wounds.

But they don't recommend that we disassemble the powder keg to render it harmless.

Alternative proteins disassemble the powder keg: Of the seven most likely causes of the next pandemic, a shift to alternative proteins would eliminate two and mitigate four more:

- They eliminate demand for animal protein from farm animals and industrial animal farming.
- They free up land for biodiversity and natural resource preservation.
- They eliminate the transport of live animals, often over thousands of miles.
- They tighten supply chains rather than expanding them.
- They slash climate emissions.

Let's address pandemic risk. Let's destroy the powder keg.

PART TWO

Changing the Meat, Not Human Nature

CHAPTER FIVE

Humanity's Favorite Food

All normal people love meat. If I went to a barbeque and there was no meat, I would say "Yo Goober! Where's the meat?" I'm trying to impress people here, Lisa. You don't win friends with salad.

— Homer Simpson[1]

Fast Food Species

When I was in high school, I ate either a McDonald's Big Mac and fries or KFC chicken for lunch—*every single day*. So did all my friends. Sitting here almost 40 years later, I can still picture the McDonald's and the KFC, but I can't remember whether Norman High School—go Tigers!— even had a cafeteria. I mean, it must have, right?

Fast forward, and I still love fast food. I don't eat it as much now, because I try to eat healthfully, but it's not exactly a sacrifice if I must "settle" for Burger King or Taco Bell because they're the only restaurants nearby and open. Add fast casual to the mix, and I'm even easier to please: I could eat nothing but Chipotle for a month.

I'm not alone. About 80% of middle-aged Americans report eating fast food at least once per week, and nearly a quarter eat it three or more times a week.[2] Income doesn't shift things much: Even among the top 10% of earners, three-quarters eat fast food weekly. I'm reminded of President Bill Clinton's frequent visits to McDonald's when he was president. A reporter asked him about it in a way that implied that it was some kind of "man of the people" stunt. Clinton was incredulous. "I just really love Big Macs," he protested.

Add in younger Americans, and the numbers climb: More than a third of all Americans eat fast food every single day—including almost half of adults aged 20 to 39.[3] Across the board, more than 10% of all calories consumed in the United States come from fast food.

I'm not surprised. During the two years I taught 11th grade in inner city Baltimore, I watched a steady stream of teachers and students eat Popeye's or Burger King pretty much daily. That included freshly minted teachers from Teach For America: graduates from elite universities, placed in some of the country's most under-resourced schools. Even these civic-minded do-gooders were eating Popeye's or BK multiple times a week. I'm not sure we had a proper grocery store within three miles of my school, but there was a BK next door and a Popeye's less than half a mile in either direction.[4*]

Why do people love fast food? Because it's quick and delicious.[5] That's it. Only a fifth of Americans think fast food is healthy, and two-thirds know it's not—even among those who eat a lot of it.[6]

Here's my basic thesis: If we're going to address the world's insatiable craving for animal meat, we're going to have to replace like for like. And

---

* It surprised me, though it should not have, to discover that fast food restaurants are disproportionately located near schools.

human beings really love meat. Fast food is exhibit A that efforts on behalf of meat reduction are going to be tough, until we can produce products that fast food outlets are excited to add to their menus. As I'm writing this, the only fast-food entrees available to meat avoiders in the United States are Taco Bell bean burritos, Burger King's Impossible Whopper, and cheese pizza.

None of the other national burger chains (McDonald's, Wendy's, Jack-in-the-Box), none of the national chicken chains (Chick-fil-A, KFC, Popeye's), and neither of the other two top fast-food restaurants (DQ and Arby's) offers a single plant-based entree. Wendy's homestyle potatoes are pretty good, and everyplace has French fries, but if you're looking for a main course, you're out of luck at the vast majority of fast-food restaurants.

America (and the World) Runs on Meat

For the past 25 years, there have been a steady stream of articles claiming that Americans are eating less meat. For example, a 2018 study from Johns Hopkins University showed that two-thirds of US consumers said they were eating less meat.[7] There are limitless similar examples.

It didn't work out that way: Per capita meat consumption in the United States in 2018 was the highest it had ever been. Until 2019. And then 2020. And then 2021. And so on.

In March and April 2025, there was a twist on the theme, a claim that Americans had been cutting back on meat, but no more!

"America Is Done Pretending About Meat," declared the *Atlantic* headline.[8] "Meat Is Back," the *New York Times* proclaimed less than a month later.[9] Both stories report that the decade leading up to 2025 had been a decade of meat avoidance. But 2025 was going to reverse that trend; it was going to be America's year of meats.

But meat was not back; it never left. In fact, meat consumption had been rising steadily in the United States since 2014 and globally since the advent of industrial animal farming in the 1940s. There was no slowdown in the decade leading up to 2025. The six highest years for per capita meat consumption in the United States? The six years leading up to these two articles.[10] And to reiterate, we're talking about *per capita* meat consumption.

Total meat production is up even more, and that's expected to continue: USDA's agricultural economists predict a steady increase in total production of beef, pork, and poultry, year-by-year, through at least 2033—everything but turkey will be setting record after record.[11] The same is true globally.

Why? As noted in the introduction, nine of Americans' 10 favorite entrees are meat-based, the one exception being grilled cheese sandwiches. The rest of the world may pick different entrees, but the love of meat is universal. In short, most humans crave meat.

The Education Fallacy

Just about every week, I have a conversation that goes something like this: I explain that GFI works to make plant-based and cultivated meat as delicious and affordable as conventional meat, because that's what we believe it will take to shift global meat production. And the response I get is some version of this: "Wouldn't it be better to advocate for a plant-based diet and help people understand how delicious and nutritious plant-based eating can be?"

Here's the thinking: Meat is bad for you. Steak is loaded with saturated fat and cholesterol. Chicken isn't the health food people think it is. Plant-based eating can be delicious! If people only knew, of course they'd stop eating meat. "I love chana masala!" I'll be told. "Who needs burgers?"

This makes sense to me; it's precisely what I thought for about 30 years. But what I've come to realize is that I'm not my audience, and most of my audience—i.e., humanity—really loves meat. Nine of Americans' top 10 entrees are meat; they like chicken and steak as much as chocolate and ice cream. Americans are not anomalous: Except for a very few countries in western Europe, meat consumption is as high as it's ever been everywhere in the world, and it's rising inexorably. And all this after decade upon decade of efforts to convince consumers to eat less of it.

Many of these conversations are with people who have cut back on meat or stopped eating it entirely. Many would not set foot in a KFC or McDonald's other than to use the restroom on a long drive. That's not most Americans, so it's useful to remember that for most people, meat is not a cause for concern; it's what's for dinner. If nine of your 10 favorite foods are meat and you're eating fast food as often as most Americans, how enthused are you going to be to discuss the kind of lifestyle change that meat avoidance would entail?

I encourage my interlocutors to think by analogy: How much less do they drive today than they did last year or the year before that or 10 years ago? How much less do they turn on the lights or use their washing machine? In many instances, the answers are, if we're being honest with ourselves: "not much" and "not much." No one ever replies that they've stopped riding in cars or consuming energy altogether. And yet we all know about the impact of both cars and energy consumption.

But many of us have shifted to an electric vehicle or installed solar panels on our homes.

For most of the world, the analogy is spot on: Eating less meat seems about as appealing as consuming less energy or driving less. Energy consumption, miles driven, and meat consumed are all rising inexorably. Global meat consumption is now more than double what it was in 1987, the year I adopted a plant-based diet. In the United States, it's up nearly 80%.

MEAT

Primary Energy Consumption (TWh) vs. Year

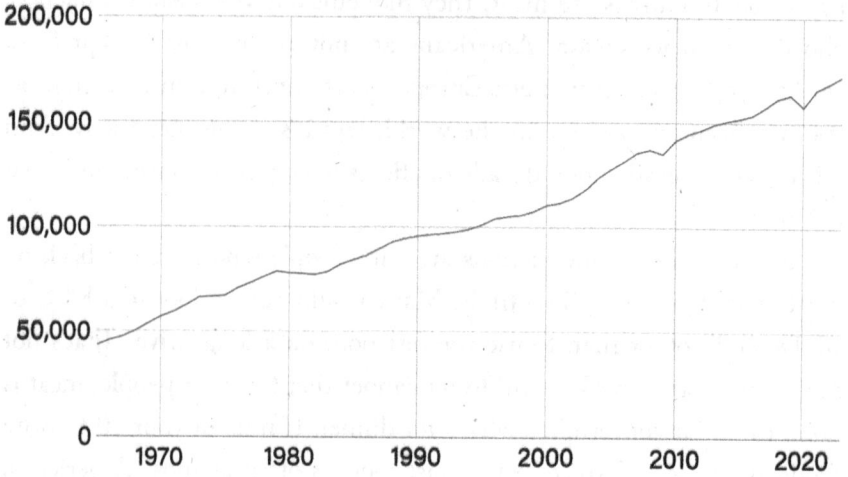

Meat Production (Tonnes) vs. Year

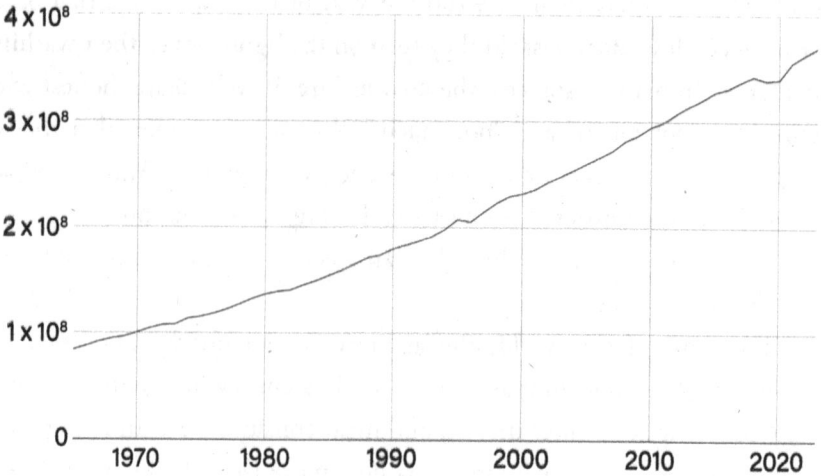

Meat: Not the New Tobacco

One of the greatest triumphs in public health is the massive reduction in tobacco use of the past 60 years, so pretty much everyone who is trying to convince consumers to change their behavior takes great inspiration from the success of that work. Often, I hear from food systems advocates who wonder if we can implement the lessons of tobacco education to change meat consumption.

I understand the appeal of the comparison, but I don't think there's much chance that conventional meat will follow the tobacco trajectory. First, the global scientific consensus that smoking will kill you has been pretty much total for more than 60 years, and that has led to herculean efforts to convince people to quit. In 1964, the Surgeon General declared that smoking causes cancer and is the primary cause of chronic bronchitis. Congress put warning labels on cigarettes in 1965 and banned cigarette advertising on radio and television in 1971. In 1998, 46 US states received $200 billion in a tobacco settlement; much of that money was funneled into anti-tobacco advertising.[12]

At smoking's peak in the mid-1960s, just over 40% of American adults smoked, and most people knew that smoking was likely to kill you, which is why 60% of Americans didn't do it. In response to the medical and political tobacco-will-kill-you consensus, tobacco use fell to 30% by 2000 and 20% by 2020.[13] The world is now awash in anti-tobacco messaging, and in developed countries smoking is prohibited in almost all indoor public spaces.

Now meat: About 98% of Americans eat meat, and you won't find a single mainstream medical body that believes that moderate meat consumption is unhealthy. Most people in the United States and around the world believe that meat is healthy, and more than half believe it's essential to good health.[14] Meat eating is ubiquitous in public and in every form of media.

Winning Hearts and Minds (but Not Arresting Meat's Upward Global Trajectory)

To be clear, public education and advocacy efforts have been wildly successful at getting attention and have improved the diets of many millions of people; that's incredibly valuable and worthwhile. What they have not done is changed the global upward trajectory of meat production and consumption, despite spectacular success at message penetration. So I'm dubious that what has not shifted the trajectory so far might do so now, since I doubt we're going to come up with some new strategy that outperforms the incredibly successful strategies of the past 30 years. Let's take a quick spin through the success of this kind of advocacy in raising awareness.

How Not to Die

In the realm of human health, there's been a steady stream of bestselling books and peer-reviewed science that tout the health benefits of eating less or no meat, going back to at least the 1990 bestseller, *Dr. Dean Ornish's Program for Reversing Heart Disease*. The subtitle notes that Ornish's recommendations are "scientifically proven to reverse heart disease without drugs or surgery"—it's the only diet that can do that. Three years later, Dr. Ornish released his bestseller *Eat More, Weigh Less*, his "program for losing weight safely while eating abundantly." Both books have sold millions of copies, and both books make an ironclad health case for eating a plant-based diet: for your heart and weight, two things most people care quite a lot about.

Ornish's findings have been published in all the most prestigious medical journals in the world, including repeatedly in the *Journal of the American Medical Association*, *The Lancet*, *Proceedings of the National Academy of Sciences*, *The New England Journal of Medicine*, the *American Journal of Cardiology, The Lancet Oncology,* and more. His work has been profiled on *NOVA* and on Bill Moyers's *Healing & The Mind,* as well as in cover stories for both *TIME* and *Newsweek,* reaching many millions of readers. The Ornish diet was rated "#1 for heart health" by *US News & World Report* for eight years in a row. Ornish gave three TED talks that racked up an average of more than 2.2 million views each and counting. He also famously turned Bill Clinton vegan. He was recognized by *Life* magazine as "one of the 50 most influential members of his generation."

Since at least the release of Dr. Ornish's 1990 heart health book, there has been a nonstop stream of peer-reviewed articles, celebrities, and documentaries touting the health benefits of plant-based eating. In recent years, a few of my favorites include the wildly successful *How Not to Die* and *How Not to Diet* by my favorite doctor—Michael Greger—and the wildly successful documentary, *The Game Changers* (producers include James Cameron, Arnold Schwarzenegger, Jackie Chan, and Lewis Hamilton).

This work has also netted support for plant-based diets from the American Heart Association, the American Cancer Society, and the Academy of Nutrition and Dietetics, which says about plant-based diets that they reduce one's risk of "heart disease, type 2 diabetes, hypertension, certain types of cancer, and obesity."

In a country where heart disease and cancer are our two biggest killers—killing north of 700,000 and 600,000 people per year, respectively—and obesity and diabetes rates just keep going up and up, you'd think that this level of proof that changing your diet can be more effective than drugs at reversing heart disease and maintaining a healthy weight would lead to a decrease in meat consumption. You'd be wrong.

While I have no doubt that meat consumption would be even higher absent all this impressive activity on behalf of meat reduction for health, it clearly hasn't been enough to decrease overall meat consumption—not even in the United States, let alone globally. Meat consumption is up 65% in the United States and has more than doubled globally since Dr. Ornish released his "we can cure heart disease with diet" book in 1990.

You Can't Be a Meat-Eating Environmentalist!

There's a similar story to be told where the environment is concerned. "Livestock's Long Shadow" was released by the UN Food and Agriculture Organization in 2006, and it made a massive splash, with stories in the *New York Times, Washington Post*, and all the world's most established media. It even netted a *New York Times* editorial. From at least 2006 to today, there has been a steady stream of peer-reviewed and popular press articles about the harm to the environment from meat production. Mark Bittman's 2007 TED talk on this theme, which has been viewed more than five million times, essentially reprised "Livestock's Long Shadow" and compares the meat industry to nuclear war and the Holocaust.[15] I'm not exaggerating.

When Al Gore and the IPCC won the Nobel Peace Prize in 2007 "for their efforts to build up and disseminate greater knowledge about man-made climate change, and to lay the foundations for the measures that are needed to counteract such change," both Gore and the IPCC weighed in on the value of cutting meat consumption for climate mitigation. The head of the IPCC from 2002 to 2015, Raj Pachauri, noted repeatedly that cutting your meat consumption in half would do more for the climate than cutting your car use in half. And when Al Gore's team pulled together "Live Earth" concerts all over the world, the Live Earth Global Warming Survival Handbook declared that the best thing any individual can do for the climate is to eat less meat.

Indeed, we've seen environmental organization after organization making clear the climate and other environmental benefits of diet shift, from the Union of Concerned Scientists to the World Wildlife Federation to the World Resources Institute. We've also seen a steady stream of peer-reviewed articles in the top science journals in the world; I cited a lot of this information in chapter 2, so I won't reprise it here.

The entreaties aren't landing, and most environmentalists eat meat, just like most of us drive cars, fly in airplanes, and turn on the lights, even if the energy mix of our local power company comes predominantly from fossil fuels. Most environmentalists see "I'm not going to eat meat," as akin to "I'm not going to use electricity."

Eating Animals!

What about animal cruelty?

Let's face it, industrial farms and slaughterhouses are not pleasant places, and pretty much everyone knows it. But for anyone who's at all unclear on the concept, Peter Singer's *Animal Liberation*, Jonathan Safran Foer's *Eating Animals*, David Foster Wallace's *Consider the Lobster*, Melanie Joy's *Why We Love Dogs, Eat Pigs, and Wear Cows*, and a steady stream of *New York Times* columns from Nicholas Kristof and undercover investigations from animal protection groups have made clear that modern farms and slaughterhouses treat farm animals badly.

And this isn't a left-right issue. In 2003, the book *Dominion: The Power of Man, the Suffering of Animals, and the Call to Mercy* offered a powerful account of industrial animal farming's unconscionable cruelty to animals. The author was Matthew Scully, who had been a top speechwriter for George W. Bush, both as governor of Texas and in the White House. When John McCain was looking for a speechwriter for Sarah Palin, he chose Scully, who wrote Palin's first national speech. In *Dominion*, Scully declared that eating meat involves siding with the strong against the weak, which is the opposite of how God and conservative principles would have us behave.

The book made a huge impact among Republicans. George Will called Scully "the most interesting conservative you have never heard of." Will recounts some of the extreme and standard abuses of industrial farms documented by Scully, as well as Scully's argument that "there cannot be any intrinsic difference of worth between a puppy and a pig," concluding that "yes, of course: You don't want to think about this. Who does? But do your duty: read his book."

Want to aim higher? Popes Benedict XVI and Francis denounced industrial animal farming, with cat-loving conservative Benedict XVI declaring that the "industrial use of creatures"—he flagged tiny cages for chickens and tiny crates for pigs—is a violation of "the relationship of mutuality that comes across in the Bible." Pope Francis, who took his papal name from the patron saint of animals, went much, much further, declaring that "the Bible has no place for a tyrannical anthropocentrism unconcerned for other creatures." It's notable that these two icons of the Catholic right and left, respectively, agreed that industrial animal farming is a sin.

While most people would agree that causing animals to suffer needlessly is immoral, very few people extend that observation all the way to the dinner table. In the *Atlantic* essay about the ostensible 2025 meat renaissance, staff writer Yasmin Tayag notes that "the thought of suffering cows releasing methane bombs into the atmosphere pains me, but I love a medium-rare porterhouse." In *Clean Meat*, Paul Shapiro's book about the early days of cultivated meat, he shares that rationalist Richard Dawkins believes that eating meat may be one of the great moral crimes of our generation, on par with slavery. And yet Dawkins eats it anyway.

On the other side of the political and faith spectrum from Dawkins, the influential conservative politico Charles Krauthammer once made that exact same point in the most storied publication in all of conservatism, *National Review.* Krauthammer suggested that within a few generations, eating animal meat would be looked at with an outraged incredulity similar to our generation's feelings about slavery. And yet he also continued to do it.

Why? I think Tayag nails it: She notes that no more than 2% of Americans avoid meat entirely, regardless of what they say to pollsters. And then she explains that "the commonality of [eating meat] can feel like a free pass." She closes with a quote from Princeton ethicist Peter Singer: "Most people can easily continue doing something they believe is wrong, as long as they have plenty of company."

I want to be clear: Meat consumption would be much higher now if not for the health, environmental, and animal protection efforts on behalf of plant-based eating. This work is valuable and important, and as you'll see in the next two chapters, it inspired the work of almost everyone who is currently working on alternative proteins, including me. But as a strategy to change meat's steady upward trajectory globally? That feels unlikely, until we have alternative meats that compete on price and taste.*

Meat Reduction Advocacy That Delivers

I'm in favor of all meat reduction advocacy, but there's one strategy in particular that I think has tremendous potential: working with institutions (e.g., public schools, universities, hospitals, food service companies, large corporations, and so on) to decrease the amount of meat they use. What's great about this kind of work is that one decision from a school or hospital administrator or corporate executive can result in a significant shift in what's served. Generally, these strategies use nudges, things like changing which entrees are promoted, changing default dishes, and so on. The nonprofit organization Greener By Default and WRI's Better Buying Lab are doing great work in this respect, as are the Tilt Collective, the Plant Based Foods Institute, and Humane World for Animals.

But even these strategies have their limitations, the biggest being that they're hard to scale and that their best-case scenario is still limited. If an entire school system or grocery chain leans in on nudges, that will do a lot of good. But it doesn't necessarily mean that another school system or grocery chain will follow. That's going to be tough to scale in one country,

---

* And nutrition, but that's essentially a given for plant-based and cultivated meat, as I will discuss in the next two chapters.

let alone globally. Also, the best-case scenario will be a decrease in conventional meat served through that institution by some percentage. Alternative meats can become the default production method, replacing all or almost all industrial animal meat over time. More on this in chapters 8 through 11.

I want to be clear: These campaigns are important. For one thing, they deliver results almost immediately. But we should be clear about what they can and cannot deliver. We need all the tools in our toolkit.

———

What I hope is clear from this discussion is that the health, environmental, and animal protection reasons to eat less meat are extremely convincing, and advocates have done a truly inspired job of making the information known. If anyone is unaware of them, it's not for lack of effort on the part of advocates. They have gotten hundreds of millions of dollars (or more) of earned media, endorsements from top movie stars to top athletes to celebrity doctors, and vegan cookbooks from here to Mars and back.

And yet the amount of meat consumed in the United States has never been higher. And globally, it just keeps going up, up, up.

With energy, driving, and meat, counting on people to consume less can influence many millions of people, and that's valuable. It's also immediate. But the past 50 years shows convincingly that these efforts are unlikely to reverse the world's massive upward energy, driving, and meat consumption trajectories, at least if consumers don't have a replacement they like just as much.

Humanity's Favorite Food

I've spent a lot of time thinking about why human beings are so resistant to dietary change, even when almost everyone would tell you that they care about clean air and clean water, preventing cruelty to animals, and living more healthfully. Why do so many people say they plan to eat less meat—and then eat more of it? What on earth is going on? Are people lying to pollsters?

No. I think the discrepancy is explained by what behavioral economist Daniel Kahneman called system one and system two thinking: System one thinking is instinctual and includes things like breathing and eating—instincts, our reptilian brain. System two has to do with considered decision-making, which includes ethics and morality.

Daniel Kahneman's most famous book, *Thinking Fast and Slow*, explains that system one thinking often prompts human actions that don't align with our values and beliefs; these decisions happen quickly and without conscious thought.

One of my favorite examples of this is a Stanford study where researchers labeled green beans four different ways: "green beans," "light 'n' low-carb," "healthy energy-boosting," and "sweet sizzlin' green beans and crispy shallots." In the Green Bean Study, the label that called the green beans sweet and sizzling and the shallots crispy outperformed all the others. And the basic "green beans" did better than the labels that stressed lightness and healthfulness.

The researchers speculated that health-focused labels send a subliminal signal that the products would not be as calorically dense, that they would be lower in fat and sugar, and that—as a result—they wouldn't satisfy our biological cravings.[16] Another thing about this study: Pretty much no one would predict that these labels would have much impact on their decision. If anything, their rational brain would have gone for the healthy green beans. That's the conflict of system one and system two.

Consider: Almost everyone would like to maintain a healthy weight, and we all know how. And yet globally, the percentage of adults who are overweight and obese just keeps rising. Between 1990 and 2022, the percentage of adults globally who are overweight grew from 25% to 43%, and the prevalence of obesity more than doubled to 16%.[17] The percentage of

Americans who are overweight and obese has been climbing even faster, reaching 73% and 42%, respectively, in 2018.[18]

For meat consumption and weight loss, it's system two thinking that is answering the pollster's question: Yes, I plan to eat less meat (or lose weight). But it's system one that actually makes food decisions for most of us.

Put another way, people don't eat meat because they've rationally weighed the costs and benefits. They eat it because it's delicious. I wasn't kidding when I put "humanity's favorite food" in the title of this book. I'm convinced that that's true. And I'm also convinced it's biological, not learned. In short, humans are wired to crave calorie-dense foods, which historically improved our odds of survival.

And of course, meat is one of the most calorie-dense and umami-packed foods we've ever encountered. Journalist Marta Zaraska captures this beautifully in her book *Meathooked: The History and Science of Our 2.5-Million-Year Obsession with Meat*. She shows how our species' long evolutionary relationship with meat, dating back to when our ancestors first started scavenging and then hunting, has left a deep imprint on our cravings; humanity's love of meat is biological and physiological.

Layer culture and repetition on top of biology and the pull becomes stronger still. We are social eaters. We eat what our friends eat, what commercials show us, what's on every restaurant menu and in every family memory. When meat is everywhere, it feels natural, normal, acceptable. That's the point that the *Atlantic*'s Yasmin Tayag was making.

Which brings me back to Homer Simpson: "You don't win friends with salad." That's not just a joke. It's an accurate depiction of most of humanity today.

What About Milk?

It is sometimes pointed out to me that plant-based milk has been growing in popularity every year, and none of the plant-based milks taste much like animal-based milk. Isn't that proof that mimicking the taste of meat is unnecessary and that we just need to produce something delicious?*

No. Consumers love the taste of meat, and they (very much) don't love the taste of milk. The only dairy product in the list of Americans' favorite foods is grilled cheese, and almost no one is ordering milk when they visit a fast-food restaurant. You personally know people who, if you ask them to tell you about their love of steak, bacon, or friend chicken, will wax eloquent for 20 minutes or more, regaling you with story after story of culinary indulgence and glee. Maybe you're one of these people. No one feels that way about milk.†

Indeed, animal-based milk has been losing in blind taste tests to soy, almond, and oat milk since at least the year 2000.[19] As of 2025, that had literally never happened for plant-based meat, as I will discuss more in chapter seven. For the average consumer, switching from cow's milk to plant milk doesn't feel like a sacrifice—at worst, it's a lateral move.

Oh, and one last point: More than two-thirds of humanity is lactose intolerant, so consuming milk literally makes them sick.[20]

Alt Meats: A Climate and Global Health Lifeline

My point is this: If we're going to address the external costs of modern meat production, we're going to want to get real about why most people make the food choices they make. And if our theory of change relies on lots and lots of people deciding they don't really like meat after all . . . well, that's unlikely to be a winning strategy, at least at the scale we need.

There's a famous Mark Twain quote that, like so many Mark Twain

---

* Because cheese production requires so much liquid milk (at least 5 liters/kilogram), global dairy production has set a new record every year since 1996, with all projections indicating a continued upward trajectory in coming decades. See: "Milk Production, 1961 to 2023," Our World In Data, https://ourworldindata.org/grapher/milk-production-tonnes?tab=chart&country=USA~OWID.

† Some people feel that way about cheese, which is why cheese consumption just keeps rising.

quotes, is both brilliant and apocryphal: "It ain't what you don't know that gets you into trouble. It's what you know for sure that just ain't so." I think of this quote a lot, because when I tell people about GFI, they often think our primary goal is to convince consumers to eat plant-based meat—right now.

But while we're enthusiastic about the early adopters and love the work of groups like Greener By Default and the Tilt Collective, our primary goal is a world where people choose plant-based and cultivated meat because they are taste-identical and similarly priced with conventional meat. So that's our focus: working with governments, scientists, industry, and innovators to create plant-based and cultivated meat that compete on taste, price, and nutrition. That's why GFI is a global network of science think tanks and not a global network of trade groups.

Taste and price parity are not the story's end, of course, but they are an essential start. Nutrition and food safety are the third- and fourth-most-important factors for consumers, but they're also baked in, which is to say: The plant-based meats that aim to replicate meat have a better nutrition profile than meat, and cultivated meat has the same profile, but it's far safer due to much lower rates of bacterial and other forms of contamination. More on these topics in the next two chapters.

A Fast-Food Species Needs a Fast-Food Solution

The bottom line: Just like the world is going to consume more energy and drive more miles, the world is going to eat more meat. Meat is just too visceral a desire, too culturally embedded, too normatively accepted. We're very unlikely to change any of this. But we can change the way meat is made.

To reiterate, I'm not suggesting that we stop our education and advocacy efforts; we've been doing that for at least 50 years, and we'll continue. But alt meats can make that advocacy more effective. Give people an easy, satisfying swap, and the ethical and environmental arguments land with more force. We need both.

More personally, advocacy is why I'm here: Reading *Diet for a Small Planet* in college didn't just change my diet; it set me on the path that eventually led me to launch GFI. That's true of many GFI team members and of most of the pioneers of plant-based and cultivated meat. Most came to this work because of health, environmental, or animal concerns.

It's also essential for helping early-stage alternative protein companies reach consumers. Until these products reach price and taste parity with conventional meat, they'll be bought by consumers motivated by ethics, health, or novelty. When cultivated meat has gone on sale, long waiting lists and lines around the block have followed, despite very high prices. That kind of demand creates valuable signaling.

The complement is also true: Alternative meats help advocacy succeed. Even at higher price points, they make it easier for people to align their meals with their values. Plant-based meat is already a billion-dollar industry in the United States and a multi-billion-dollar industry globally. As the products improve and become less expensive, sales will grow. And once plant-based and cultivated meat are price-competitive and delicious, advocacy will become even more effective, because consumers will have compelling side-by-side options.

But without a viable alternative to conventional meat that competes on price and taste, it's hard to see how the current upward meat consumption trajectory could possibly change: What's the new strategy? Alternatively, what causes us to think that the old strategy will reverse meat's upward trajectory now, after decades of not doing so?

No Silver Bullets

People often tell me two things that I want to address briefly: that alt meats are not a silver bullet and that there's more to food than whether it's delicious and affordable. It seems I'm so enthusiastic about the potential of alt meats to help solve the huge and intractable problems detailed in part one that in my alt meats evangelical fervor, I give the impression that I think the opposite. I don't.

Of course alt proteins are not a silver bullet, and I don't think I've ever heard anyone claim that they are. I certainly haven't.* Food systems are complex, and so too hunger, climate change, and global health. The world is in desperate need of "all of the above" thinking and action across the range of complex issues discussed in chapters 1 through 4. I'm in favor of all of it.

So too renewable energy and electric vehicles: Are these silver bullets? Of course not. As I noted in chapter 2, we also want to encourage conservation and efficiency: We need more energy-efficient light bulbs and buildings and everything else. We also need more walkable cities and great public transit and bullet trains. Plus, we'll need plans to ensure that building out a renewable grid is done carefully and that battery components are sourced responsibly. All of the above.

Alternative meats also don't address every problematic aspect of global food production. But as I'll discuss more in chapter 10, by freeing up vast tracts of land, they make it far more feasible to generate the kind of food systems reset the world so badly needs.

They also appear to be our best shot at changing the massive upward trajectory of industrial meat production globally; nothing else seems to have any reasonable likelihood of doing more than decreasing the massive upward slope. That's important, since all the problems discussed in part one get worse if the volume of meat consumed continues to rise.

To be clear, my goal with this book is to convince you that alt proteins are an essential intervention, not that they are the only intervention worth pursuing.

---

* Do a search for "alternative proteins" and "silver bullet," and everything that comes back will be a critique of the idea that alt proteins are a silver bullet.

Necessary, but Not Sufficient

Food choices are complex, of course. A lot more is involved than how delicious and affordable something is. Food is about culture and family and so many other things. Again, yes of course. No one is suggesting otherwise.

But what we've seen is that most consumers aren't going to choose plant-based or cultivated meat unless it's at least as delicious and similarly priced. Until we have those two metrics cleared, none of these other factors will matter much.

More important: Once we have those two metrics cleared, consumers will be able to factor in additional concerns: the food safety benefits, the nutrition benefits, the environmental and other benefits discussed in part one of this book. And at that point, the products will have better profit margins than conventionally produced meat, and those profit margins will just keep widening. So the private sector incentive to tout the other benefits and respond to any concerns skyrockets.

But unless we create price- and taste-competitive alternative meats, conventional meat consumption will just keep rising, and with it all the harms detailed in part one of this book.

———

America is a fast-food nation, and humans are a fast-food species. Most of us crave the fat and protein that meat offers, the dense calories, and the umami flavor. That's why global meat consumption has been going up for as long as records have been kept and is expected to continue rising for as long as anyone has forecasted.

Yes, some people are satisfied with plants, and yes, some people will change their diets for health or environmental reasons or out of concern for animal protection. Some people will also embrace "small is beautiful" by avoiding air conditioning, consuming less, walking and biking more, and so on.

But just like energy consumption is as high as it has ever been and just going higher, so is meat consumption. We need a strategy that works with human nature rather than against it. Creating taste- and cost-competitive plant-based and cultivated meats is going to be one essential tool in our toolkit.

CHAPTER SIX

From Science Fiction
to Science Fact

Sometimes a new technology comes along, and it has the capability
to transform how we view our world . . . If what you're doing is not
seen by some people as science fiction, it's probably not transforma-
tive enough.

— Sergey Brin[1]

It tastes like chicken.

— Suzy Welch

Sergey Brin's $300,000 Hamburger

In August 2013, a team of scientists unveiled the world's first cultivated
hamburger at a high-profile tasting event in London. The burger had been
grown by Dutch scientist Mark Post, a tissue engineer, medical doctor,

and former professor of medicine at Harvard Medical School. The effort was bankrolled by Google co-founder Sergey Brin, at a cost of $300,000.*

I remember hearing about it at the time and thinking: Cool, I guess. But also, $300,000 for a burger? That's not going to do much good, at least not in my lifetime.

When I started working on what became GFI in the fall of 2015, vanishingly few people had considered the idea that animal meat could be produced without the need for live animals, and most of those who had considered it didn't expect commercialization for decades, if ever. The project was exploratory.

I was no different: I was aware of the concept, but as a near-term consumer product, it felt about as viable as flying cars. One of the first questions I considered between October 2015, when I started to conceptualize GFI, and June 2016, when I started to hire staff, was whether we should work on this "cultivated meat." It seemed intriguing, but it didn't seem like something that might change the way human beings had been producing meat for 12,000 years, at least not any time soon.

The first cultivated meat company in the world, Upside Foods, was founded in October 2015 (the same month I started working on GFI) by Uma Valeti, a Mayo Clinic–trained cardiologist who was teaching medicine at the University of Minnesota. Back then, it was called Memphis Meats. Like Mark Post, Uma talked about making the exact same meat—but cleaner. What they meant by "cleaner" was compelling: Cultivated meat could have far less bacterial contamination, none of the drug residues common in conventional meat, and—for seafood—no mercury, dioxins, or microplastics.

---

* Mark Post is widely considered to be the father of cultivated meat, because he was working on the technology before almost anyone else. Post taught medicine and tissue engineering at Harvard Medical School for six years and Dartmouth Medical School for eight. Subsequently, he served as chair of the biophysics and physiology departments at Maastricht University in the Netherlands. In 2016, he co-founded one of the first few cultivated meat companies, Mosa Meat. Post's influence had a lot to do with the Dutch government's 2022 investment of €60 million into the science of cultivated meat.

As a quick aside: This book makes the case for plant-based and cultivated meat, but it's not a history. If you'd like to learn about Dutch researcher Willem van Eelen, who put cultivated meat onto Mark Post's radar and spent much of his life working to see his dream turned into a reality, as well as the full stories of his daughter Ira van Eelen, Mark Post, Uma Valeti, Josh Tetrick, Isha Datar, and other cultivated meat pioneers, I recommend *Clean Meat* by Paul Shapiro and *Billion Dollar Burger* by Chase Purdy.

To refresh: Cultivated meat is actual animal meat, grown in cultivators that look like brewing tanks rather than inside animals. Just as a seed or cutting can grow into a full plant, a sesame-seed-sized sample of animal meat can grow into vastly more animal meat. It's not a coincidence that the two pioneers of cultivated meat are both medical doctors. They'd already grown human tissue in labs. The idea of growing meat using the same fundamental process—but with food-grade nutrients and scaled for food-grade production—didn't feel to them like a giant leap. It is pretty much the same science, applied differently.

Or at least that was the claim.

I'll admit, I wasn't immediately convinced. Growing meat in cultivators instead of animals sounded possible in theory, but possible and promising are two very different concepts. The idea seemed like a long shot. Still, I was curious. I wanted to understand whether cultivated meat could realistically be produced at any kind of scale. And if it could, how soon?

My earliest mentor into the ways of science startups was Ryan Bethencourt, co-founder of an investment fund called IndieBio, which is focused on scientific innovation. IndieBio was Upside's first investor. Ryan told me that when he finished reading Upside's application, it took him about 30 seconds to decide that he wanted to fund the company. As with all venture investments, this was 70% how impressed he was with co-founder Uma

Valeti and 30% how convinced he was of Uma's vision. Mostly, he figured that Uma would figure out the best path forward, even if it wasn't this precise idea. But the more he talked with Uma and read about the idea of growing animal meat without the vast inefficiency and climate emissions of conventional meat production, the more excited he became. He saw environmental impact, and he saw dollar signs.

Bethencourt's background is in science, and mine (very much) is not, so pretty much every sentence coming out of his mouth befuddled me, other than this: He told me that cultivated meat was an idea whose time had come. In venture capital, many investors expect a high failure rate, but if one in 10 companies hits big, they more than compensate for the failures. Ryan told me that in his view, Upside was the investment that would pay for the rest. And then some. His enthusiasm is nicely captured in the excellent documentary film about Upside and Uma: *Meat the Future.*[2]

I love Ryan, then and now, but I also suspected that he expected every IndieBio investment to be the one that paid for all the rest. The man's enthusiasm is both real and infectious, but it's also total. About everything. So I remained dubious: If what Ryan was saying was true, how was this the first time I was hearing from anyone that cultivated meat production might be the basis of a for-profit company, not in 20 years but right now? The consensus as I understood it was that if cultivated meat was going to be viable, we'd know in a decade or two.

"You've got to meet Uma," Ryan replied. "He'll convince you."

Unlike Ryan, whose genuine ebullience about the future of the world was unmitigated, Uma's focus was as calculated as you'd expect from an Indian-born medical school professor who had trained in cardiology at the Mayo Clinic.* After Mayo, Uma spent six years as a cardiologist at the St. Paul Heart Clinic, three years as president of the American College

---

* At the *Wall Street Journal*'s Global Food Forum in 2017, I asked Cargill vice president Joe Stone what had convinced America's second-largest meat company to invest in cultivated meat. "Have you met Uma Valeti?" he asked rhetorically. He went on to rave about Uma's intelligence and integrity and Cargill's desire for partnerships with people they both trust and like.

of Cardiology, and five years as a professor of cardiology at the University of Minnesota. While he was teaching, he also served as president of the Twin Cities board of directors for the American Heart Association.

And then he decided he wanted to make meat.

I started my first meeting with Uma thinking that his company was highly unlikely to succeed, and I ended wanting to read everything I could find on the concept of cultivating meat, to pressure test what Uma was telling me. After spending a lot more time with Mark Post and attending his first conference, spending a lot more time with Uma and his cofounders, and reading the limited number of journal articles that existed at the time, I was convinced.

Putting Cultivated Meat Science to the Test

While I was convinced, I also had sufficient humility to know that I wasn't a great judge of scientific viability.

In June 2016, GFI hired our first six team members. One of them was senior scientist Liz Specht, who five years later would become our first vice president for science and technology. Her first charge was to critically assess what we were hearing from Ryan, Uma, Mark, and others. In short: How ready for commercialization was cultivated meat, really?

Liz dived into the research and spoke with scientists from all over the world. She reminded me in 2025 that she had started this analysis highly skeptical. She had worked with mammalian cells, and she found them "finicky, expensive, and woefully easy to contaminate." She liked the idea of supporting the science of cultivated meat, but she didn't like the idea of companies trying to figure out how to make a profit in the short term. The science was too nascent.

But she devoured the science that Mark and Uma shared with her. Tissue engineering for human medicine has been making progress for decades, and much of that work has direct application to cultivated meat.

We are not starting from zero. They also pointed her to biopharma and introduced her to biopharma scientists in the private sector. After all this research, her mind had changed; she emerged optimistic.

The challenge, she found, is exactly what Mark and Uma said: Take what we know from tissue engineering for human medicine, swap in food-grade ingredients and manufacturing systems, and scale.

Easier said than done, of course: Media was absurdly expensive and contained ingredients that we knew couldn't scale. We would have to figure out replacements. Plus, no one had ever produced tissue at anything like the scale that would be necessary for cultivated meat production. There were many more challenges. But multiple well-funded scientific disciplines already held the expertise needed, and the scientific questions were clear.

GFI scientist Erin Rees Clayton puts it this way: "Tissue engineers learn most if not all of the skills needed to produce cultivated meat. They know how to grow cells, and they understand media and scaffolding. While academic scientists may not work with bioreactors directly, they do understand the kinds of factors that influence cell growth and differentiation. So they have at least a basic knowledge of what a bioreactor would need to do."

There's nothing inherent to the process, Liz concluded, that precludes cultivated meat from becoming cost-competitive with conventional chicken. Post and Valeti, both medical doctors, believed that the efficiency gains from growing meat directly, without the costs of keeping animals alive or supporting bones, feathers, and skin should allow lower costs at scale. It made sense, and the science seemed like it *might* back up their assessments.

In early 2017, Liz completed her first full cultivated meat analysis. While it didn't include a complete assessment of technological readiness levels, it made clear that four areas were critical: cell lines, media to feed the cells, scaffolding for the cells to grow on, and bioreactors for them to grow in. And the biggest hurdles? Lowering media costs and producing at volumes that make cost-competitive production feasible.

If cultivated meat had not appeared to be commercially viable, that would have been fine: We would've doubled down on our original project, building the strongest possible ecosystem to help plant-based meat compete with industrial animal meat.

Flash forward almost a decade: Between January and June 2025, I engaged with 27 of the world's top cultivated meat scientists, asking them whether they believed that cultivated meat could reach price parity with conventional meat and what they saw as the key hurdles. Media and cultivator scaling were the two hurdles almost everyone raised. On the first, there was universal optimism: Researchers described extensive progress on media cost reductions and optimism that costs would keep coming down.

On the second, some felt that cost-competitive premium-priced products could be produced right now, if the resources for building factories were available. Others thought that bioreactor design was going to take time, iteration, and money. Everyone was enthusiastic about the possibilities of artificial intelligence for speeding progress.

Multiple scientists also mentioned their enthusiasm for cultivated meat as a plant-based meat ingredient. This would allow the creation of entirely alt meat products as hybrids, helping plant-based products to clear their final taste hurdles and cultivated meat to scale, so that costs could come down.

Liz's analysis didn't just help shape GFI's strategy; it helped shape the broader industry. Vow (an Australian cultivated meat company) founder George Peppou told me that Liz's analysis of cell-culture media, and the viability of slashing costs that she outlined, was critical to his decision to found the company. Quite a few founders and investors have shared similar aha moments that came from Liz's early cost modeling and other aspects of GFI's work.

When Upside Foods was raising its first major investment round, Liz presented her findings to prospective venture investors and incumbent meat companies, to offer third-party validation for the basic science. Following strong initial meetings and extensive due diligence processes with the Memphis Meats team, several of these investors ultimately participated, including industry giants Cargill and Tyson Foods.

The round was led by Steve Jurvetson, founder and managing director of Future Ventures, a tech forward investment VC based in Silicon Valley that invests in companies that they believe have transformative potential for all of humanity. For example, Steve was an early investor in Tesla and SpaceX, where he has served on the board since 2009.

Steve eats meat at most meals and loves bacon, but he's also incredibly uncomfortable with meat's impact on animals and our environment. He had been looking for a viable cultivated meat company for years. When Uma came calling, he was ready. "I saw the inevitability of manufacturing meat without animals long before I found a compelling way to get there," Steve told me. "I can't imagine harvesting animals for slaughter 100 years from now. And certainly not on Mars."

Steve personally appealed to Bill Gates and Richard Branson, convincing both to join him among Upside's earliest investors.

Two other proof points came from McKinsey: In late 2016 and again in early 2019, their clients asked them to investigate cultivated meat. While I was encouraged by Liz's analysis, I can still remember how happy I was to receive the late 2016 email from Markus Gstöttner from McKinsey's London office, asking if GFI could introduce him to the chief scientists for all of the cultivated meat companies that existed at the time, so that he could prepare an analysis of the industry's prospects. At this point, GFI consisted of just nine full-time staff, and we were just getting started.

Wow, I thought, this is basically a *pro bono* McKinsey analyst who will check the cultivated meat science and tell us what we're missing.

Liz shared her research and scientific analysis with Markus, who also joined us on a field trip to Israel in 2017 to meet with Israeli biomedical engineering professor Koby Nahmias from the Hebrew University of Jerusalem and Shulamit Levenberg, who was dean of the Israel Institute of Technology's biomedical engineering department. Of the first eight cultivated meat companies, three were in Israel, and Koby and Shulamit were co-founders and chief science officers of two of them, Believer Meats and Aleph Farms.

Markus's assessment? He quit McKinsey and co-founded one of the most successful European cultivated meat companies, Meatable, in early 2018.

The second analyst, Yossi Quint from their Manhattan office, was already a cultivated meat enthusiast when he joined McKinsey, and much to the consultancy's credit, it gave him the time and resources to interrogate his enthusiasm. Like Markus, he liked what he was learning. In 2021, he left McKinsey to start Ark Biotech, a company focused entirely on bioreactor development for cultivated meat.

I don't know how often an analyst at McKinsey researches a brand-new industry and ends up leaving to found a startup in that space—but for cultivated meat, we're two for two.

Momentum, Milestones, and the March of Science

In recent years, my optimism for the cultivated meat endeavor has grown. Part of that is based on my many conversations with top cultivated meat scientists and global partners from the World Resources Institute, UN Foundation, Bezos Earth Fund, and others. But there's a lot of external validation as well.

In 2020, the National Science Foundation awarded $3.55 million to the University of California at Davis to launch a Cultivated Meat Consortium focused on stem cell amplification, tissue differentiation, and scalable bioprocessing. A year later, Tufts University received $10 million from the USDA to launch the National Institute for Cellular Agriculture. Over the next few years, more than $30 million was granted to cultivated and plant-based meat research from the USDA alone.*

Additional government-funded research centers dedicated to cultivated meat followed quickly in Canada, China, Denmark, Israel, South Korea, and the UK, with additional major investments from India, Japan, the Netherlands, Singapore, South Korea, Australia, New Zealand and multiple additional European nations, as well as from the European Commission.[3] Every one of these rounds of funding from governments for cultivated meat research centers or cultivated meat scientific research moves the entire endeavor forward.

None of these centers and none of this funding existed at the beginning of 2020.

What's driving this global wave of research funding? It's a very important coming together of scientists and government funding agencies: The science of cultivated meat requires both government support and a deep bench of scientists. Governments want to fund established and successful scientists. Those scientists want to feel confident that if they submit grant proposals, they are likely to receive funding. At GFI, we've worked hard to create enthusiasm on both sides of that equation—and it's working.

All over the world, top-tier scientists, from tissue engineers to cell biologists to mechanical engineers, are working on cultivated meat across both the public and private sectors. GFI's science team has endless stories

---

* UC Davis principal investigator David Block told me that GFI scientist Liz Specht's visit led to the university's decision to work on cultivated meat, and Tufts principal investigator David Kaplan told me that GFI's scientific ecosystem building led to their USDA grant.

of key researchers all over the world convincing their program managers in government agencies that cultivated meat should be a funding priority.

For me, the caliber of scientists dedicating their careers to this space is one of the most encouraging signs of its viability. These are people already in the upper echelon of the scientific establishment, and they have limitless options. They're choosing cultivated meat because they see the incredible potential for the world, and they see a viable path forward for the technology.

Take David Kaplan, who was serving as Chair of the Department of Biomedical Engineering at Tufts, with more than 1,000 peer-reviewed papers under his belt, when New Harvest fellow Natalie Rubio brought the idea of cultivated meat to him in 2016. Since then, he's gone all-in, overseeing graduate students, writing papers, and launching the aforementioned National Institute for Cellular Agriculture.

At Tufts' annual conference on cellular agriculture in 2024, I asked him how he was feeling about the field's trajectory. "In the last three to five years," he told me, "the progress on cell isolation, cell immortalization, media cost reduction, and many other factors gives me complete confidence that we're going to get to where we need to go."[4] Kaplan's cultivated meat lab has grown from one to more than 20 full-time researchers in less than five years, bolstered by USDA and Massachusetts state government grants.

David's confidence is echoed by others. Eric Schulze is a molecular biologist and chief technology officer at the cultivated meat company Omeat. Eric spent six years at the Food and Drug Administration evaluating novel biotechnologies before joining Upside Foods and then Omeat, which was founded in 2021 by Ali Khademhosseini, a former professor at both MIT and Harvard Medical School, and one of the world's most renowned tissue engineers.[5] In response to my early 2025 survey, Eric replied: "If cultivated meat weren't feasible, researchers would already know. There's no evidence we can't invent new ways to lower costs. If anything, you could almost calibrate a watch to the pace of breakthroughs."

Mark Post agrees, telling me that he's been impressed by the plummeting production costs for cultivated meat, especially considering how little has been spent on open-access research. When he unleashed the Brin burger on the world, we really didn't know whether cost-competitive cultivated meat would ever be possible, he said. But now we know. "Tremendous advances have been made," he explained.

That enthusiasm is showing up in the data. Between 2013 and 2020, a total of 27 peer-reviewed papers focused on the science of cultivated meat were published. In 2021, there were 36. In 2022, that number doubled to 72. In 2023, the number jumped again to 110. And in 2024, we hit 160.[6]

One of the most influential studies came from Professor Nahmias from Israel. Published in *Nature Food* in late 2022, the study shattered multiple assumptions about cell densities, media costs, and production efficiency.[7] GFI lead scientist for cultivated meat Elliot Swartz told me that Koby's study was the most comprehensive study on cultivated meat production published to date.

One thing I really love about the cultivated meat field is the collaboration. Scientists like Koby and Mark are working with private companies, but they see the cultivated meat endeavor as so much larger than their personal interests, and so they share their progress with the entire world. They are working to build not just one company but an entire field. I find this incredibly inspiring.

In 2024, *Nature Biotechnology* reached out to discuss an editorial. After researching the technological and environmental impacts of cultivated meat, the editors made the case that governments should help to ensure its success, noting that "with better coordination between public and private efforts, cultivated meat will both do well and do good." The editorial explained: "In the best-case future, cultivated meat will contribute meaningfully to emissions reductions and food security, help end the cruelty of industrial meat farms, and free up land devoted to livestock for biodiversity restoration, carbon sequestration, small farmers and agroecological farming."

After the piece ran, deputy editor Kathy Aschheim told me that governments have been critical to innovation in many areas of daily life, including a range of applications in biotechnology, especially in the early days of exploration. She shared her view that cultivated meat looks scientifically promising and that all innovation takes time. Governments should help the sector to succeed. "Governments have unique levers to boost R&D in areas of national interest, helping startups cross the 'valley of death' to successful commercialization," she told me. More on these topics in chapters 9 and 10.

Patents are another sign of progress. Between the launch of the Brin Burger in 2013 and the first eight companies being founded in 2015 and 2016, the entire field saw just 10 patents issued. Between 2017 and 2024, that number exploded to more than 1,000.[8] This includes work from foundational companies like Upside Foods, Mosa Meat, Aleph Farms, and Believer Meats—each one founded by medical doctors or PhDs in bioengineering (or both). Patent activity captures just a sliver of private sector scientific progress, since many scientific and technical advancements are protected not as patents but as trade secrets.[9]

Will Anyone Eat It?

Of course, strong science doesn't matter much if no one is willing to eat it, and one of the most common questions we hear at GFI is whether the "ick factor" will kill cultivated meat in its cradle. The concern is that consumers might not like the idea of meat produced in factories rather than naturally, from live animals.

I'm not worried, in part because consumer surveys related to cultivated meat are extremely promising. For example, a global survey by Barclays of 5,000 consumers found that two-thirds would purchase cultivated meat if it were available, and half said they'd pay more for it.[10] Unsurprisingly, younger consumers were more enthusiastic: Almost 80% of 18-to-34-year-olds said they'd eat it, compared to roughly one-third of respondents over 55. Results were similar around the world, with the strongest enthusiasm in China.

That's consistent with a broader body of research. In a review of 40 peer-reviewed consumer studies on cultivated meat, Chris Bryant and his team at the University of Bath found several recurring patterns: Most people haven't heard of cultivated meat, and many have questions, but most are willing to try it. And many say they'd be happy to eat it regularly.[11] Even when it's framed in less appetizing terms—e.g., "lab-grown" or "in vitro"—consumer openness holds.

A March 2024 survey by Purdue University reinforced this point: two-thirds of US consumers said they'd be willing to try cultivated chicken and beef, which the survey referred to as lab-grown, though of course it will be grown in a factory, not a lab.[12]

And what about the consumers who decline? Among those who say they're not interested in cultivated meat, the top concerns are palatability and price, even when the prompt includes price and taste parity explicitly. In the Purdue study, respondents rated the expected palatability of conventional chicken and beef as 55% and 63% higher, respectively, than cultivated versions; many participants said that they expected cultivated meat to be mushy or have an odd texture. So it's even more remarkable that two-thirds were still willing to try it.[13]

All of this bodes well. The cultivated meat products coming to market taste as good or better than conventional meat (stories forthcoming!). They'll also be safer, with lower risk of chemical residues and bacterial contamination. As more people become familiar with the idea and have the chance to taste the products, consumer enthusiasm should only grow.

Selling the benefits of cultivated production may also help: In one study, behavioral scientists from Harvard and the University of Pennsylvania asked consumers to imagine a burger that "tastes, looks, and smells exactly like the usual beef hamburgers" but is "made in a factory by growing the meat cells from cows," with "much less environmental impact" and "no cows killed." Consumers were willing to pay more than twice as much for the cultivated burger. When the same burger was described as made from the cells of "world-famous Wagyu beef" and promised to "taste better" than any conventional beef, consumers said they'd pay 2.5 times more.[14]

Of course, we should take these results with a grain of salt. Consumers are notoriously bad at understanding or predicting how they make food choices. In this case, though, that observation may bode well for cultivated meat, since there are at least three reasons to think that these polls are underselling eventual consumer demand.

First, new foods always take time to gain traction. Human beings are wired to be cautious with unfamiliar food, which is an evolutionary instinct to avoid poisoning. Those concerns will fade as more and more cultivated meat is produced and eaten.

Second, there's no moral pressure in these surveys. We're not asking people whether they exercise or donate to charity. There's no social desirability bias pushing people to say they'll eat cultivated meat. Most people love conventional meat. The fact that so many are open to cultivated meat even before it exists to any meaningful degree is encouraging.

Third, taste is the clincher. Researchers at Singapore Management University found that enthusiasm for cultivated meat increased significantly after consumers tried it, because it tasted good. Consumers who had doubts came away impressed by the flavor, and so their attitudes shifted dramatically.[15]

It's also worth noting a few other things:

First, researchers consistently find the most enthusiasm for cultivated meat among the participants who eat the most meat. And second, men are more excited by the idea of eating cultivated meat than women, an observation that researchers attribute to the general tendency of men to be more open to unfamiliar foods than women.[16] My strong hypothesis is that women and casual meat consumers will become more and more enthusiastic about cultivated meat once it's on sale, becomes more familiar, and the benefits become better known.

Chris Bryant from the University of Bath pointed out to me that since it's the heavy meat consumers who are most excited about cultivated meat, the already-promising numbers are even more promising, since it's the people who are eating a disproportionate amount of conventional meat who are most excited to replace it with cultivated meat.

"It Tastes Like Chicken"

Among GFI's first handful of significant supporters were legendary former General Electric CEO Jack Welch and his wife Suzy, a professor at NYU's Stern School of Business. Suzy has been a close friend for many years, and she's also been my counselor and confidant from the beginning of GFI. I thought she might serve in a similar role for Uma Valeti and his work to turn Upside into the next great American meat company. I also just adore both Uma and Suzy, and I thought they should know one another.

So that's how it came to be that Suzy, Uma, and I were standing in Jack and Suzy's kitchen in October 2016, hearing about Uma's heart surgeon-to-meat-baron trajectory, as Uma cooked up some cultivated chicken. I've heard quite a few vegetarians talk about their first time eating cultivated meat, and many of them talk about their reservations. Not Suzy and me. We could not have been more excited!

Suzy took a bite and, eyes beaming, declared, "Wow, it tastes like chicken!"

Uma smiled and replied, "It is chicken."

Both Suzy and I were totally blown away. I have not eaten conventional meat in quite a while, but even I could tell how different this was from the plant-based chicken products that existed in 2016—how much better.

It was at that point that I realized something that felt then and still feels fascinating: This cultivated meat endeavor is both remarkable—bordering on science fiction—but also mundane: Chicken that tastes like chicken. How surprising is that, really?

Seven years later, a small group of team members from GFI US and GFI APAC were touring Good Meat's cultivated meat facility in Singapore, alongside Bezos Earth Fund future food director Andy Jarvis. We had the chance to eat some cultivated chicken, and Andy's reaction was visceral: His eyes lit up, he smiled, and he declared spontaneously: "FUCK ME!" as Good Meat founder and CEO Josh Tetrick looked on, beaming.

At that point, I grabbed my phone and started recording: "Oh my God!" Andy continued, "it's impossible to tell the difference. It's better. That is amazing. No way!" He was laughing and smiling the whole time.

Andy told us afterward that he'd been dubious about cultivated meat: Like the Purdue study participants, he just didn't think it would taste very good, and he thought our focus should be on perfecting plant-based meat. But tasting was believing. From then to now, many of us at GFI will sometimes refer to cultivated chicken as "fuck me chicken," especially in the context of first-time tasting experiences.

Clean Meat

There's another reason to believe that cultivated meat should be able to transcend any concerns that consumers have about it being unnatural: comparing its production and safety to the production and safety of conventionally produced meat.

Remember back in the introduction when I mentioned that every university on the planet that has an agriculture department has meat science courses and labs? So do all the big meat companies. So 99.99% of laboratory work with meat happens with conventional meat, not cultivated meat, making it simply incorrect to refer to cultivated meat as "lab grown."

What's happening in those meat labs? They're working on how to reduce bacterial counts, extend shelf life with additives and preservatives, find chlorine levels that can kill pathogens on chicken without harming consumers, and address the tumors, hernias, and other unpleasant realities of modern meat production.

Wait, what: tumors? So . . . cancer meat?

Sometimes, yes. Generally, the animals are too young for cancer to have developed, so more often tumors that can develop into cancer. But also yes: in some cases, cancer meat.

During the Biden administration, the poultry industry asked USDA to eliminate multiple requirements related to inspection for *avian leukosis*, a viral disease that causes tumors in chickens, and to rescind the ban on the sale of meat from chickens with tumors. The administration complied, rescinding multiple inspection requirements and allowing the sale of birds with tumors, including cancerous tumors, as long as the tumors are cut out. If cancer is visibly obvious, the affected organ is also condemned.[17]

Similar regulations have applied to cattle and pigs for many years: Cut out the tumor and sell the meat, unless there's visible evidence of cancer that has spread beyond the tumor.[18] Yum.

I will spare you a description of umbilical and groin hernias in pigs and the special slaughterhouses that exist for pigs with protruding intestines to decrease the likelihood the organs will rupture and spew the contents of the animals' intestines all over their carcasses.[19] Um, I guess I won't, actually, but I will leave it there, other than to note that these hernias are common and that meat from these animals is on sale all over America.

What about bacteria on meat? Salmonella and campylobacter poisoning, for instance, overwhelmingly come from contaminated poultry, due to what *Consumer Reports* called the "filthy and crowded conditions in which most chickens are raised," as well as the USDA's limited authority over what happens on farms.

Even in slaughterhouses and meat plants, where the USDA does have oversight, the agency's tolerance for contamination is shockingly high: They don't intervene on salmonella unless almost 10% of whole chickens,

15% of chicken parts, or 25% of ground chicken test positive.[20] The thresholds for other meats are similarly lenient.

The CDC's advice about how to ensure that your meat is safe to eat? Cook the crap out of it. Literally. They want us to heat chicken to 165°F and treat it like toxic waste: Don't wash it (splash contamination), don't cross-contaminate, and scrub your hands and kitchen surfaces like you're doing surgery.[21] In other words, treat your kitchen like a biohazard lab.

Finally and perhaps most alarmingly, USDA regulations allow more than 100 different drugs in meat if they stay under what's called the "Acceptable Daily Intake."[22] The drugs that USDA approves for human consumption via contaminated meat consumption include Amoxicillin, Ivermectin, Progesterone, Ractopamine, Testosterone, and Tetracycline.[23]

I don't know about you, but the precise amount of Ractopamine that I would like in my meat is zero.

USDA begs to differ.

And even with these shockingly lax guidelines, enforcement isn't exactly airtight. The USDA routinely finds meat that exceeds legal limits. When that happens, they issue Drug Residue Warning Letters.[24] A USDA audit in 2024 found that 20% of meat labeled "raised without antibiotics" was contaminated with antibiotics.[25] I wonder what antibiotic residue numbers look like for the overwhelming percentage of meat that doesn't carry the antibiotics-free label.

I'm reminded of a *New York Times* column by Nicholas Kristof about several Johns Hopkins University studies suggesting that poultry on factory farms are routinely fed the active ingredients in Tylenol and Benadryl, banned antibiotics, and even arsenic. Kristof noted that poultry-industry literature recommends Benadryl to reduce anxiety among chickens, since stressed birds grow more slowly and develop tougher meat. Tylenol and Prozac, he reasoned, probably serve the same purpose. He ended the column wondering if a Prozac-laced chicken nugget might ease his growing depression about what's happened to farming.[26]

All of this is domestic land-based meat. The picture gets even more concerning when we consider seafood. As noted in chapter two, the US imports more than 90% of the seafood we eat, much of it from countries with lax food safety standards. The FDA doesn't have the budget or the workforce to inspect more than a fraction of what comes in.* According to a Government Accountability Office report to Congress, imported fish often have high rates of bacterial infection. To counter that, fish farmers use antibiotics and antifungal agents, which can leave behind drug residues.

The FDA does have a protocol for what they call "significant violations" of food safety law, which is supposed to include warning letters and follow-up inspections within six months. But of 125 such violations reviewed by GAO, the agency followed up on just 11% within that timeframe. Almost half were never followed up on at all.

And as bad as all this is, developing economies have it even worse. Peer-reviewed literature on the overuse of antibiotics in meat from emerging economies is both deep and grim. As countries rapidly industrialize poultry and pig farming, the use of feed additives and antibiotics—including many that are banned in the United States—has exploded. In some places, they're both legal and endemic. In India, for example, a study found that consuming just 60 grams of chicken per day—that's less than four McNuggets—could lead to toxic levels of arsenic ingestion.[27] That's one example among many.

Poisoned Seafood: Mercury and Microplastics

Perhaps the scariest example of meat contamination is mercury in seafood, which can have very quick adverse impacts on brain function. I'm

---

* *Imported Seafood Safety* (GAO-21-231, 2021); *Food Safety: FDA Should Strengthen Inspection Efforts* (GAO-25-107571, 2025) (from 2018–2023, FDA's best year for overseas food facility inspections was shy of 10% of what the Food Safety Modernization Act requires).

still surprised by how few people know that human mercury exposure is almost entirely a function of fish consumption and that most fish contain toxic mercury.[28]

I'm reminded of a front-page *Wall Street Journal* article about fifth-grader Matthew Davis, who went from doing well in school and sports to struggling to add numbers, catch a football, or hit a baseball.[29] His parents spent thousands of dollars on tutors, to no avail. Eventually, Matt was diagnosed with a learning disability. But that wasn't it at all: Matt had been eating three to six ounces of tuna per day, which caused his blood mercury levels to spike to double what the EPA considers safe. The *Journal* noted that some tuna samples contain seven times more mercury than tuna's average rate of contamination.

The toxic effects of fish consumption can start in the womb: Fish consumption during pregnancy is directly linked to learning disabilities in children. As *New York Times* reporter Hiroko Tabuchi explained, "Most people with mercury in their bodies get it from eating contaminated seafood, and, even in small amounts, it can harm the brains of unborn children and have toxic effects on the human nervous, digestive, and immune systems."

The EPA estimates that mercury exposure in utero may increase the risk of learning disabilities for more than 75,000 US newborns each year.[30] The loss of intelligence in children attributed to the exposure of pregnant women to mercury-contaminated seafood is estimated to cost the United States billions of dollars of lost productivity every year.[31]

That bears repeating: Contaminated seafood is the primary source of human mercury exposure. It's already making Americans sick. And it may be compromising the brain development of tens of thousands of babies annually.

And unfortunately, the problem isn't going away any time soon. In February 2024, Tabuchi reported on a study that analyzed mercury levels in tuna from 1971 to 2022. A French research team found that mercury concentrations have remained stubbornly high over those five decades, with no signs of improvement.

By contrast, while cultivated meat is actual animal meat, it won't be contaminated with pathogens, drug residues, or environmental toxins. And cultivated seafood will have all the healthy fish fats but none of the mercury, dioxins, microplastics, or heavy metals. All of this is true for plant-based meat too: It contains none of the contaminants that are routine in animal-based meat.

What began as a one-off publicity event—a quarter-million-euro burger cooked up at a London press conference—has become a growing global science project.

Mark Post and Uma Valeti were the early visionaries. Now, their work is part of a full-blown ecosystem of scientists and policymakers working to make safer, cleaner, climate-friendly animal meat a reality.

What once seemed like science fiction is moving very quickly toward scientific fact.

The main question now is whether we'll give it the support it needs to change the world.

In chapters 9 and 10, I'll explain why I think we will.

CHAPTER SEVEN

The Most Important Scientific Problem in the World

In the next few decades I believe that [cultivated] and plant-based meat will become the norm and animals will no longer need to be killed en masse for food . . . Meat production has not changed in 10,000 years. It's time that changed.

— Richard Branson[1]

"You Don't Actually Like Food, Do You?": The Early Days of Veggie Meat

At the beginning of 2015, I was where most people were on the issue of plant-based meat: I hadn't thought that much about it, or at least not about it having any kind of transformative potential. I'd eaten what passed for plant-based meat since 1987, the year I read *Diet for a Small Planet* and changed my diet, but I'd never really thought about plant-based meat as a supply-side intervention.

In 1987, veggie burgers were heavy on the veggie and light on the burger; they were a veggie mush, shaped kinda like a burger. If you asked for a Veggie Big Mac or a Veggie Whopper, you got everything but the burger: special sauce, lettuce, cheese, pickles, onion, on a sesame seed bun. It wasn't too appetizing. At some point in the early 1990s, I think it was, the products got better. We had Boca Burgers, Yves veggie dogs, and in 1996, Tofurky.

They weren't mush, but they also weren't trying to replicate animal meat. They were low in fat, high in protein, zero cholesterol, and tucked into the health food aisle, often in neglected corners of the grocery store. Finding that aisle felt like a scavenger hunt. And when you did find it, it was often unclear that management knew the area was there. "Has anyone ever brought a mop back here?"

For decades, I brought the latest vegetarian meats to friends and family, really leaning in on their protein content. They were much higher in protein (as a percent of calories) than conventional meat; that's what made them more meat-like than the products that came before. But they were also almost totally devoid of fat. That's what made them unappealing to most people who enjoyed meat.

My friends and family humored me, saying things like "this isn't horrible." One summer in Minnesota, I brought veggie dogs to a cookout, only to learn they turned rubbery on the grill. My cousins and I ended up batting them over the garage with a baseball bat. They flew surprisingly well.

My friend Gene Weingarten, the two-time Pulitzer-winning *Washington Post* humorist, and I visited Johnny Rockets to try their Boca Burger–based Streamliner when it first came out in the early 2000s. "It tastes a lot like meat," I assured him. He couldn't finish it. "You don't actually like food, do you?" he asked. I have a lot of stories like this one.

Our bodies are designed to crave fat and flavor, and in those days, veggie dogs and veggie burgers contained less than half the calories and about a fifth as much fat as animal-based dogs and burgers. For example, the Boca burgers used by Johnny Rockets clocked in at 110 calories and two grams of fat, against 290 calories and nine grams of fat for a standard burger.

Like many vegetarians, I had trouble putting myself in the shoes—palates?—of a meat-eater. But it really should have occurred to me that something with 110 calories was no substitute for something with 300. It turns out most people aren't looking for a tasteless veggie burger that's lower in fat. They're looking for something that mimics animal meat.

Ethan Brown and Pat Brown—no relation—understood this from the start.

Ethan and Pat were the first plant-based meat entrepreneurs who were focused on competing with animal-based meat. For the first many decades of veggie dogs, veggie burgers, and veggie nuggets, the products were vegetarian first, and meat . . . well, they weren't really meat at all. That wasn't the goal; the products were designed for health-conscious consumers who were avoiding or cutting back on meat.

How Climate Change Revolutionized Veggie Burgers

In 2009, two men with the same last name had the same radical idea: What if plants could be used to make meat? Not veggie burgers. Not tofu dogs. But food that meat lovers would actually want to eat. That's when Ethan Brown launched Beyond Meat and Pat Brown started working on what would become Impossible Foods.

For both men, it was the massive climate footprint of animal-based meat that inspired their desire to create a like-for-like replacement. Ethan earned a public policy degree from the University of Maryland and spent more than a decade deep in the renewable energy sector, first working on grid infrastructure and then clean hydrogen. Surrounded by people trying to save the world from climate disaster, Ethan noticed something that he found odd: They were all still eating meat. Some didn't know the impact meat had on the climate. But even those who knew didn't change their diets.

Ethan asked what he considered to be a simple question: If energy transition was a worthy goal, what about protein transition? The energy transition was focused on replacing fossil fuels with renewables. What about a protein transition that replaced industrial animal meat with plant-based meat? After all, meat is just protein, fat, minerals, water, and trace nutrients. Plants have all those things. What if you could rebuild meat, piece by piece, using plants?

He searched the patent office and scientific journals, looking for anyone else working on the same project. He found two University of Missouri scientists, Fu-hung Hsieh and Harold Huff, who had received a USDA grant to experiment with plant-based meat textures. From their collaboration, Beyond Meat was born.

Pat Brown was a scientific heavyweight by 2009: an MD and PhD from the University of Chicago, tenured chemical engineering professor at Stanford, cancer researcher, a pioneer in genome mapping. He also co-founded the Public Library of Science, one of the earliest and boldest efforts to make scientific knowledge free and open to all. Reading his CV, one would be hard-pressed to think of someone less likely to become a corporate executive, let alone a corporate executive with designs to build the largest meat company in the history of the world.

In 2009, Pat took a sabbatical to focus on climate change, and a friend said something that would change his life: "If you could make a burger that McDonald's would serve instead of a burger from a cow, then that would be the fastest way to solve the problem." As Pat later explained, "It was a throwaway comment, but I realized that was exactly it."[2] He also realized, as a scientist, that this should be possible. He was shocked to discover that as far as he could tell, no one had ever tried.

So Pat launched Impossible Foods in 2011. His goal: not just to make a decent veggie burger, but to decode the molecular magic of meat and recreate it from the ground up.

Ethan was thinking similarly. I remember talking with him early in 2011 or 2012, so very early in the trajectory of his company. He told me that Beyond Meat was hiring chemists to mimic the umami flavor of meat and tissue engineers to figure out how plant proteins could be manipulated to behave like animal proteins. "It's not going to be easy," Ethan explained, "but we already have products that are more meat-like than anything else on the market. And we're just getting started."

It was Ethan and Pat's visions that inspired me to start GFI. They weren't trying to convince people to eat differently. They were trying to give people the exact same meat experience they already loved, just made with a tiny fraction of the impact on our climate and forests.

Plant-Based Meat for Everyone

Pat incorporated Impossible Foods in 2011, and after five years and tens of millions of dollars' worth of scientific research, the Impossible Burger debuted at two very posh restaurants: David Chang's Momofuku Nishi in New York City and Traci des Jardins' Jardinière in San Francisco. It was kind of like launching the Prius at a monster truck rally. Chang had previously declared vegetarians unwelcome at his restaurants, even adding chicken stock to the rice to eliminate the one animal-free item on the menu. Des Jardins, similarly, had nothing vegetarian on offer.

That was the point. Impossible wasn't marketing to vegetarians. They were taking the fight to the heart of meat culture. And it worked. The high-end focus was a PR coup: Global coverage of the launch was off the charts. "The plant-based burger that bleeds," was one of the more common headlines.

That pro-meat and pro-science ethos continues to guide both companies. The Impossible Foods homepage declares: "Calling All Scientists!" Click on the link, and you read: "Let's build meat that's more delicious, healthier and more affordable than today's obsolescent animal-derived products."

Beyond Meat echoes Impossible's focus on remaking meat: "Why are we spending so much money on lithium-ion batteries for cars and so little money on creating plant-based meat?" Ethan Brown asked rhetorically.

Taste is King

They're right to focus on replicating the taste people love. When it comes to food, taste is king, and we've already established that the vast majority of humans love meat.

Let's try a quick exercise: Raise your hand—mentally, if you're in public—if you like Ben & Jerry's ice cream.

I use this exercise all the time. I'll show a slide with eight Ben & Jerry's flavors (usually the nondairy ones, to include dairy-avoiders) and ask who loves ice cream. Without fail, every hand shoots up. People smile. There's happiness in just thinking about how great ice cream tastes.

It's no surprise. The human body is wired to crave ice cream—dense calories, fat, and sugar. Meat works the same way: calorie-dense, satisfying, delicious. That's why we love it.

My point is this: While pretty much everyone would tell you they think about health as a critical part of their food choices, almost no one thinks about the health consequences of ice cream at the moment that I'm asking the question. Decisions around food are more visceral than most of us realize.

Study after study confirms the hierarchy of food choice: Taste and price are first and second (or second and first); everything else is a distant third. *New Yorker* journalist Nicola Twilley summed it up: "In surveys, customers tell you that they want healthy choices, but analysis of purchasing patterns reveals a different hierarchy of priorities: customers care about taste above all else, and value for money to a certain extent; any other claim that a product touts, be it health benefit or environmental impact, lies far behind."

That's been a problem for plant-based meat. In every survey of consumers who have tried plant-based meat but no longer consume it, the top two reasons they've stopped are that they didn't like the taste, and the products cost too much.[3] And in surveys of why most people haven't tried plant-based meat at all? Consumers anticipate the products won't taste very good, and they can see that they cost more.[4]

The fact that taste and price are keeping consumers away should indicate that making more delicious and affordable products will increase consumer adoption. And indeed, social scientists have spent quite a lot of time asking consumers if they would purchase plant-based meat that matched conventional meat on taste and price; the numbers are encouraging: between 21% and 41% say they would.[5]

For example, a massive survey across 27 countries asked: "Assuming each tasted equally good, had equal nutritional value and cost the same, which one of the following do you prefer?" More than 40% preferred "meat-like alternatives" over "real meat from animals." In China, which has been through quite a few food safety scares in recent years, preference for plant-based was more than half.[6]

As with our discussion of cultivated meat consumer research, there's reason to believe that these numbers are soft. For example, a study published in the journal *Appetite* found that about one-third of consumers would select either plant-based or cultivated burger if it tasted and cost the same.

But here's the thing: More than 90% of the naysayers declined because they thought the alt burgers wouldn't taste good enough or would be too expensive. In other words, exactly what we found with cultivated meat consumer surveys: Participants rejected the question's premise; they simply could not fathom a delicious and affordable plant-based burger.[7]

I put my suspicion that these numbers were soft to the test. Over the course of a few weeks, I asked random strangers—in line to board a flight, Airbnb hosts, Lyft drivers, Starbucks baristas, other Starbucks customers—if they would eat plant-based meat that tasted identical to animal-based meat and cost the same or less. I found that most people said no. Why not? More than half simply didn't think that plant-based meat could ever taste as good as animal-based meat, even though I'd just asked them to assume that it did. Many of them had tried the products and didn't like them. That was that. Once I convinced them to suspend disbelief and assume my premise, enthusiasm for plant-based meat more than doubled.

I ran this little experiment twice: Once in 2023 and again in 2025, with 12 people in each "study." The results were the same.

To be clear, even 20 to 40% is great—that's a massive untapped market. Consider that current sales are roughly 1%. But the actual market is likely at least twice that. The human inability to imagine plant-based meat that competes with conventional meat on taste and price is skewing consumer research, just like it's skewing consumer research into cultivated meat acceptance. With actual taste and price parity, consumer acceptance will be even higher.

Love Meat? This Plant-Based Burger's for You!

When I tell people what GFI does, the response is often some version of "Oh, that's not for me, I love meat!" That's the "no way is this going to taste good" bias in action. These people assume our goal is to create a broader array of products for vegetarians and vegans, or at least for people who are trying to cut back.

The flip side, which I also hear regularly, is from vegetarians: They'll say something like "I don't know; I'm not really that excited about plant-based meat. I'd rather eat a bean burrito."

With meat lovers, I generally respond with something like, "well then, this is absolutely for you! We're not trying to create more options for vegetarians; what would be the point in that? We're trying to make meat that YOU as a meat lover will enjoy." That exchange has been the start of some very encouraging conversations.

With vegetarian and meat reducers, I sometimes respond tongue in cheek: "If you're a vegetarian, then we don't care what you eat!" I'm not trying to be mean, though in the early days of GFI, I did toss out the line with enough abandon that I offended some people. I try to be more careful now, but the point feels important for a mission-driven organization: There's no impact in convincing a vegetarian to switch from a plant-based burrito to a plant-based burger.

Okay, so that's not strictly true: We do need the early adopters, whether vegetarian or not, as I discussed in chapter 5. The broader point is just that the goal of plant-based meat is to win the hearts and palates of people who love animal-based meat, not to create additional options for people who weren't going to eat meat anyway.

Plant-Based Meat: Ultraprocessed Outlier

Some readers are probably thinking about all the negative attention given to ultra-processed foods (UPFs) in recent years, so I'd like to unpack what makes UPFs problematic. Although you can find articles talking about all kinds of different theories, the best science—which I will discuss below—really does boil down to this: They typically contain excessive amounts of "bad" nutrients like sugar and saturated fat, while lacking "good" ones like fiber.

There's a strong scientific consensus that saturated fat is harmful; it's linked to numerous health conditions, including the number-one killer in America: heart disease. Likewise, there's agreement that fiber is beneficial; it's associated with lower rates of heart disease, diabetes, and obesity. Most Americans eat far too much saturated fat and not nearly enough fiber; in fact, 94% of us fall short on fiber intake.[8]

Most ultra-processed foods are heavy on the bad stuff and light on the good. Think chips, soda, pastries, and ice cream: loaded with sugar, saturated fat, or both. By contrast, most unprocessed food like grains, legumes, fruits, and vegetables are rich in fiber and contain little or no saturated fat. In this sense, the shorthand critique of UPFs mostly holds up.

Against that backdrop, plant-based meats look pretty great: The most popular plant-based meats contain fewer calories, less fat and saturated fat, and have a lower caloric density than their conventional counterparts. At the same time, most of them contain a healthy dose of fiber, where animal meat has none. They also contain more protein as a percentage of calories, which is important to many consumers.[9] So on the key metrics that define why UPFs are unhealthy, most of the meatiest plant-based meats come out far ahead.

I looked at eight of the plant-based meat products that are both among the most popular in the United States and that were also in the top tier for satisfying meat enthusiasts in professional taste panels run by NECTAR, a nonprofit initiative focused on alt proteins sensory research (at least half of regular meat consumers liked these eight products as much as their animal-based counterparts). Here's what I found[10]:

- The Impossible burger, the Beyond burger, and the Morningstar steakhouse style burger: at least 20% fewer calories, about half as much fat, and about one-third more protein as a percentage of calories (compared to a conventional burger).
- The Impossible and Gardein breakfast sausages: about 40% fewer calories, less than half the fat, and two to three times as much protein as a percentage of calories (compared to conventional sausages).
- The Impossible and Morningstar nuggets: at least 15% fewer calories, about half as much fat, and a bit more protein as a percentage of calories (compared to conventional nuggets).
- The Impossible hot dog: about 20% fewer calories, half as much fat, and almost three times as much protein as a percentage of calories (compared to a conventional hot dog).

Honestly, I find it remarkable that these products, which are the closest to mimicking the precise taste of animal meat according to professional sensory panels, are so much healthier than their animal meat counterparts. I wonder if just a bit more saturated fat would push them to absolute taste parity.

Ultraprocessed Foods: Science Exonerates Plant-Based Meat

A 2019 landmark study in *Cell Metabolism*—often referred to as "The Hall Study," after lead author Kevin Hall of the NIH—sparked headlines worldwide. This was the first randomized, controlled trial comparing ultra-processed and unprocessed diets. The finding? People on the ultra-processed diet consumed more calories and gained weight.[11]

Why? Because the ultra-processed cohort ate things like potato chips, sugary drinks, processed meats, and French fries—foods high in fat and sugar and low in fiber. As Hall told the *Wall Street Journal*, "People eating the ultra-processed foods had to consume more calories to attain the same level of satisfaction and fullness . . . one way ultra-processed foods may contribute to weight gain is that they often contain more calories per gram compared with less-processed foods."[12]

Put simply: The UPFs are calorie-dense; that is, they pack in lots of fat and sugar per bite. As Hall and coauthors noted, "the ~85% higher [caloric] density of the nonbeverage foods in the ultra-processed versus unprocessed diets likely contributed to the observed excess energy intake." So this is key: Most animal-based meats are far more energy-dense than their plant-based counterparts.*

Many people mistakenly believe the Hall study controlled for macronutrients like fat, protein, and carbs. It didn't. The processed group ate 500

---

* This is true for all of the products listed a few paragraphs up from here, as noted.

more calories per day (!!!), all of which came from carbs and fat.* That's 70 more grams of empty carbs and 25–26 more grams of fat—every single day. Plus, the sugar in the ultra-processed diet came mostly from sweets (absent in the unprocessed diet), and the fat had nearly double the saturated fat ratio (34% versus 19% of fat calories), which would be expected to increase the risk of chronic disease and all-cause mortality.[13]†

Honestly, I'm not sure this study should have commanded so much attention; how surprising is it that eating an extra 3,500 calories every week will lead to weight gain?

A meta-analysis in the *British Medical Journal* echoed Hall's findings. It linked UPFs to increased risk of heart disease, diabetes, cancer, and premature death. The review highlighted these foods' "higher levels of added sugars, saturated fat, and sodium; higher energy density; and lower fiber, protein, and micronutrients." It also pointed to how their "greater energy density, faster eating rate, and hyper-palatability" make overeating more likely.[14] After all, our bodies are designed to crave fat and sugar, and ultra-processed foods tend to be high in one or both. Again, most plant-based meats do better across all these categories.

What if you control for UPFs that are high in fat and sugar? A 2023 study found that the strongest links between UPFs and chronic disease were to animal-based products and sugary beverages. This should not be surprising, since the former tend to be very high in fat and devoid of fiber, while the latter are very high in added sugar. No link at all was found for ultra-processed bread, cereals, or plant-based meat.[15]

A follow-up study published in *The Lancet* in November 2024 found

---

* The source of the confusion is that the media often reports the study this way, because the third sentence of the abstract reads, "Meals were designed to be matched for presented calories, energy density, macronutrients, sugar, sodium, and fiber," and Table 1 includes the "presented diet," in which total calories and macronutrients are basically identical. It's disappointing that the authors included a table with presented calories, fat, and sugar, but no table with actual consumption levels; that might have helped to prevent the broad misunderstanding of the study's findings.

† Plus, the omega-6 to omega-3 proportion was 11:1 in the ultra-processed diet, verses 5:1 in the unprocessed diet (see endnote for citation).

that plant-based meats and milks were actually associated with strong positive health outcomes: They appeared to cut the risk of developing diabetes in half.[16]

There is an outlier study from 2024, which generated headlines claiming that plant-based meat consumption leads to negative health outcomes.[17] *Scientific American* called foul and explained what happened: Although the study itself said no such thing, the press release for the study falsely claimed that "products intended to replace animal-based foods—such as plant-based sausages, nuggets and burgers" were linked to adverse health outcomes.[18] Not true.

There were no plant-based sausages, nuggets, or burgers included in the study; the "plant-sourced" foods that the press release writer claimed were "intended to replace animal-based foods" consisted of: pastries, buns, cakes, fries, candy bars, soft drinks, salty snacks, processed pizza, and distilled spirits. Meat alternatives were also there, listed dead last and responsible for 0.2% of total calories consumed: two out of every 1,000 calories.

Participants ate 35 times more pastries, 14 times more fries, and 10 times more soda than they did meat alternatives, which included tofu and tempeh. I'd wager that when you think "plant-sourced" or "meat replacement," you're not thinking about French fries, soft drinks, and candy bars. The study's takeaway is what every other UPF study has shown: Diets high in fat and sugar are unhealthy.

Strip away the occasional misleading headline, and you're left with this: Plant-based meat that is focused on replicating the taste of animal meat generally has more protein, less fat, saturated fat and cholesterol, more fiber, and lower caloric density than animal meat. You'd expect it to be healthier, and you'd be right.

Stanford medical school scientists tested Beyond Meat's chicken, beef, and pork against organic chicken, organic pork, and grass-fed beef. Published in the *American Journal of Clinical Nutrition*, the study found improved LDL cholesterol levels, lower saturated fat intake, and reduced heart disease risk in the plant-based group. The group also lost weight.[19]

It's true that plant-based burgers have more sodium than raw meat, but the Stanford team found no significant difference in sodium intake, even when study participants ate plant-based meat three times a day instead of organic and grass-fed meat. Why? Because conventional meat is typically salted during cooking. That's why processed meats like sausages, nuggets, and hot dogs generally have as much or more sodium than plant-based versions, and the Impossible Whopper's salt content matches that of the beef-based Whopper.

Three systematic reviews that compared plant-based and conventional meats confirm the Stanford findings.[20] When you swap a food that's higher in fat, saturated fat, cholesterol, and calories for one that has more fiber and protein, you end up with a healthier option.

But again: Healthier only matters if people make the switch. And they'll only make the switch if plant-based meat can match conventional meat on taste and price. That remains the big challenge.

Nerds Over Cattle

In April 2016, then-Google CEO Eric Schmidt took the stage at the Milken Global Conference to reflect on emerging technologies that he felt had the potential to improve life for humanity by at least a factor of 10 in the fairly near future. The science didn't have to be fully baked just yet, but everything pointed to success if sufficient resources could be mobilized.

Schmidt listed off a slate of transformative ideas: 3D-printed buildings, virtual reality, self-driving cars. To the surprise of many in the room, he also talked about alternative meats. He called the category "nerds over cattle," describing it as a technical challenge that could be surmounted: how to update the incredibly inefficient process of cycling crops through animals, making meat from plants, and growing real animal meat, no live animal required. If we could do that, he suggested, that would go a long way toward solving the most intractable cause of climate change and de-forestation: the human desire to eat meat.

The next few years saw a flurry of activity in the plant-based meat space, with Beyond Meat raising hundreds of millions of dollars and Impossible raising $2 billion. Those two companies spawned both copycats and a wave of investor enthusiasm. Analysts from Bloomberg, Credit Suisse, BCG, JP Morgan, and Kearney predicted a bright future for plant-based meat. For example, Kearney projected a $450 billion market by 2040; Credit Suisse estimated sales of $1.1 trillion by 2050.[21][22]

These reports shared a single, crucial caveat: Taste and price would make or break the category. If plant-based meat didn't taste the same or better and cost the same or less, it wasn't going anywhere. As BCG put it in one of their reports, most consumers were willing to pay exactly nothing for the health and environmental benefits of plant-based meat; hitting price and taste parity would be essential to success.

Those numbers indicated that analysts were optimistic that the companies could clear the scientific and scaling challenges. The science seemed within reach, and Beyond Meat and Impossible Foods had proven that significant investment could yield the meatiest plant-based burgers the world had ever seen. Surely scaling wouldn't be too difficult.

San Francisco, We Have a Problem

In 2018, two years after GFI's launch into the world, our scientists started thinking about the plant-based meat challenge very differently; they were beginning to realize that replicating animal meat with plants was going to be more challenging than we'd thought. The next year, the nonprofit Food System Innovations (FSI), which would later launch the NECTAR initiative, conducted the first-ever professional palatability panels, which confirmed our concerns.

No plant-based meat tested by FSI, which included all the most popular ones, satisfied the 98% of consumers who love meat, though the Impossible Burger came close. All the chicken and hot dog offerings performed ... abysmally. For the most part, the science was simply not there yet. It turns out that not many companies were even trying.

In my experience, most people—even those excited about alternative meats—believe that making plant-based meat is a culinary endeavor. Mix the right ingredients, get creative with spices and flavors, and voilà: meatless meat.

Their intuition is failing them. That's not it at all.

GFI scientist Erin Rees Clayton explained to me that plant-based meat is asking biology to do something outside its nature: Plant proteins are globular; animal proteins are fibrous. Plant oils are liquids at room temperature; animal fats are solids. Replicating the structure and functionality of meat with entirely different ingredients isn't just a matter of culinary craft; it's a scientific problem.

The two plant-based meat pioneers, Impossible and Beyond, understand this. They weren't playing with recipes. They hired tissue engineers, molecular biologists, chemists, meat scientists, extrusion engineers, plant breeders, and more. Their goal was not different in degree from the plant-based meat companies that had existed up until that point; their goal was different in kind. They were building a *brand-new* category from scratch, applying the rigors of science and engineering to food.

Erin expanded on this challenge, explaining her view that the underlying science of plant-based meat is, contrary to my intuition, a lot more complex than the science of cultivated meat: "Virtually no one is trained across the entire plant-based meat production process. Plant breeders can modify and improve crops but often don't know what happens once those crops leave the field. Protein chemists can extract high-purity proteins but may not understand how different extraction techniques affect flavor, digestibility, or food functionality. Food scientists understand formulation but may not have experience with extrusion. Meat scientists know meat, but they've rarely applied their knowledge to plant proteins."

Pat Brown put it bluntly: "The most important scientific problem in the world," he said, was "What makes meat taste delicious?" And Impossible Foods was going to find the answer. Pat recruited a team of scientists and treated plant-based meat like an Apollo-level mission.

Allen Henderson joined Impossible in 2014 and worked there for about a decade. He spent his first two years as one of many scientists working on the 2016 burger launch. He told me that most meat and food companies spend less than 1% of their budgets on research, while pharma companies often invest closer to 30%. Pat's goal, he said, was to out-science pharma. Allen holds a PhD in biochemistry and focused his doctoral and postdoctoral work on protein science. Still, "during my time at Impossible, I learned so much," he told me. "It felt like we were all living in the protein Renaissance."

The Impossible team figured out how to mitigate the off-flavors from plant proteins. Nature creates many of those off-flavors, Allen told me, specifically to protect plants from being too delicious. They don't want to be eaten. The team built a gas chromatograph-mass spectrometer to identify flavor molecules created when meat cooks. They tested heme (an iron-containing compound that contributes to the meaty taste of meat) from 31 different sources, from clover to cattle to soy, finally settling on a process that produces a synthetic soy-based heme.[23]

Even someone as deeply trained in protein sciences as Allen said there was no way around trial and error: "You really don't know what you're going to get until you try it," he told me. Scaling up or down changed everything. Small tweaks could dramatically shift texture or flavor.

The Quest to Be the Next Gardenburger

After Beyond Meat and Impossible Foods started raising substantial sums and Beyond went public at a multi-billion-dollar valuation, a flood of plant-based startups appeared, each pitching themselves as "the next Beyond

Meat" or "the next Impossible Foods." Over and over, their pitch decks featured Impossible and Beyond as comparators. And over and over, though with a few notable exceptions, it was obvious they were fooling themselves.

The biggest red flag? The R&D budgets for almost all these companies were tiny or nonexistent, they didn't have a chief science officer, and they projected product launches within six to eight months. That's possible, but only if you're not actually trying to compete with conventional meat on taste. They weren't. And they didn't.[24]

Recall that Impossible Foods—founded by one of the world's top scientists—spent north of $100 million and more than five years before releasing a product. Similarly, Beyond Meat spent tens of millions of dollars and three years on research before launching its first product. Its breakout hit, the Beyond Burger, took seven years and tens of millions more. That Beyond Burger was the only product besides the Impossible Burger that performed well in FSI's 2019 taste panels.

Another flag: expensive ingredients and clean labels. Many of these companies' pitch decks for investors would distinguish themselves from Impossible and Beyond by noting that they used healthier proteins like lupine or lentils.[25] They would also display side-by-side nutrition comparisons indicating that their products would have fewer ingredients, less fat, less sodium, and no unpronounceable ingredients. The focus on lupine and lentils guaranteed that the product would cost a lot more. The focus on low fat and clean labels guaranteed that it would taste nothing like animal meat. In other words: the health food strategy of the past four decades.

All of these veggie meat companies with no research budgets and a commitment to non-soy plant proteins, low fat, and clean labels? They were not the next Impossible; they were the next Gardenburger. That's fine; that was the entire category until Beyond and Impossible were launched. But just be clear: You're competing for a share of the $1 billion US veggie meat market; you're not ever going to compete with the $2 trillion global animal meat and seafood markets.

Pat Brown believes the deeper issue is a failure of imagination: People can't picture plants precisely mimicking animal meat. Their thinking is stuck in the era of veggie burgers and tofu dogs. He told *The New Yorker*'s Tad Friend in 2019: "Nobody else has caught on to the fact that this is the most important scientific problem in the world, so their results are just a reheated version of veggie burgers from 10 years ago—maybe with a little lipstick on them."

Breakthrough Opportunities in Plant-Based Meat

I mentioned that there are a few plant-based meat companies that are following in the science-forward footsteps of Impossible and Beyond, and so there are, though at this point they're all extremely small. It's not hard to figure out which companies these are: They are focused on scientific research, they're using soy protein or mycoprotein as their primary ingredient, and they are explicitly focused on creating plant-based meats that compete with conventional meat on deliciousness and affordability.

There are more than 150 cultivated meat companies, and there are dozens of companies working in fermentation and precision fermentation and molecular farming. In these later categories, companies are programming soy plants to produce actual dairy proteins and also creating protein out of thin air.* It's all fascinating, and I'm convinced that it's the future of food. But it befuddles me that against this food tech backdrop that finds more than 150 science-focused cultivated meat companies and dozens of science-focused precision fermentation companies, there are maybe 10 similarly science-focused plant-based meat companies: companies that are leaning in on the science of perfectly replicating animal meat with plants, but at a lower price.

The opportunities for success in plant-based meat are at least as significant as the opportunities in cultivated meat, but only if you hire a

---

* Read all about it at "Fermentation," Good Food Institute, accessed June 15, 2025, gfi.org/fermentation.

top-notch CSO and additional scientists, focus on science and engineering, and don't try to rush a product to market before you have shown in a professional sensory study that it tastes very close to or better than its animal counterpart. I'd love to see a lot more science-focused plant-based meat companies.

Better yet: I'd love to see more plant-based meat research companies that are focused on business-to-business endeavors, with no plans to launch a consumer brand at all, unless they spin it out of their research company: Designing fit-for-purpose soy, better fats, better flavoring agents, better production methods, and so on. We have dozens of cultivated meat companies that are focused on every step of the value chain. They are researching media, scaffolding, cell lines, and cultivators—and offering licensing agreements to all comers. But we have few companies with this kind of focus on the plant-based meat side, despite overwhelming opportunities that could elevate the entire sector. Yet.

And to be clear: My strong suggestion here is to launch plant-based meat companies that are not also brands. When the fermentation company Meati failed, venture capitalist Steven Finn shared an insight that strikes me as extremely smart: "There was real technology at Meati, incredible work at the bleeding edge of fermentation tech," he explained. But trying to build a consumer brand while also developing world-class technology, as Meati was doing, is nearly impossible. Companies should do one or the other. "This isn't just about Meati," he wrote. "It's a broader warning. Don't let amazing science fail because you picked the wrong go-to-market path."[26]

Plant-based meat companies, take heed.

Taste Parity Is Not (Only) Impossible

The private sector has shown us that taste parity is possible.

In 2025, NECTAR tested 122 of the most popular plant-based meat products across 14 categories, and the results were encouraging: 20 of the products performed well, with at least half of consumers liking them as much as their animal-based counterparts. The four products that did the best were the Impossible Burger, Impossible chicken breast, Impossible chicken nuggets, and Morningstar's chicken nuggets.

Impossible also led in several other categories. Its hot dog and meatballs were the only products in those categories to score well. Its breakfast sausage was one of just two to perform well, along with Gardein's. Other strong performers included the Vegetarian Butcher's breaded and unbreaded chicken fillets, Beyond Meat's burger, Redefine Meat's burger, and a few others.[27]

The report was a highlight of 2025 for me, because until the Impossible Burger launched in 2016, we didn't know for sure that it was scientifically possible to make a plant-based burger that consumers would enjoy as much as beef, and in 2018, there were no plant-based chicken or pork products that came anywhere close. Now we have both the Impossible burger and three chicken products that have almost reached parity, plus another 16 that many meat eaters liked just as much. That's real progress.

It also proves something important: Subpar products are a choice. Companies that hire scientists, invest in R&D, and conduct consumer sensory testing can make products that meat eaters enjoy. That said, it's not easy. Impossible has spent far more on research than any other company, and its board and investors have always understood it as a science-first food company—more akin to cultivated meat companies than to most of the plant-based sector.

Impossible understands the assignment in another way too: It's willing to make the ingredient choices that will be necessary to compete. For example, Impossible uses soy as its base protein, which puts them on the fast(er) track to taste and price parity. GFI lead plant-based meat scientist Nikhita Mansukhani Kogar explained to me that soy has fewer off-notes than other legumes, and it's also the least expensive plant protein by a healthy margin. And it's easier to texturize than any of the other common proteins that are used for plant-based meat, in part because the industry has been working with soy for decades. Another benefit of soy is that for consumers who are concerned about protein bioavailability, soy matches animal meat, where most of the other principal proteins for plant-based meat do not.

Impossible is also willing to fight fire with fire: The company's hot dog is the first plant-based dog that's ever come anywhere close to taste parity with a beef hot dog, because one dog contains 120 calories, seven grams of fat, and 430 milligrams of sodium per 47-gram dog. That's important if we want someone who loves hot dogs to also love plant-based hot dogs. It may feel like a lot of calories, fat, and salt, but conventional dogs contain more calories, twice the fat, one-third the protein, and the same amount of salt.

Another thing that's clear from the 36-page NECTAR report and its raw data: The more closely a plant-based product resembles animal-based meat, the more consumers like it. This aligns to what I assumed was true: that one of the most obvious reasons that most people don't like plant-based meat is their nutritional profile. The NECTAR data backs that assumption.

Plant-based meat that tastes like meat isn't a pipe dream. It's happening. Impossible has proven it repeatedly. And while burgers, dogs, and nuggets are easier than steak or salmon, we're starting to see meaningful scientific advances in protein structuring and 3D printing that could carry over into structured meats.

Public Sector Science: Solving Plant-Based Meat's Taste Challenge for Everyone

Some of the biggest challenges in plant-based meat production are these: First, the flavoring agents that are required to mask the soy flavor are incredibly expensive; plant-based meat companies don't buy in large enough volume, so they can't command lower prices. Second, although soy is the best option, it's still not a great option, because the soy protein has not been optimized for plant-based meat; again, this is a volume issue. Third, plant fats don't love to behave like animal fats; the best plant fat to stand in for animal fat is coconut oil, but even that isn't great, and it's also expensive. Finally, much of the equipment used to manufacture plant-based meat is stuck in the 1950s; sales are not sufficient to justify the kind of production innovation that could drive increased efficiency. Finally, the plant-based meat itself is produced at low volumes, which prevents companies from realizing economies of scale.*

Each of those challenges could be an excellent idea for a company, helping the entire plant-based sector to improve. They're also great options for university research, funded by governments. Imagine a world where the secret of Impossible's best products was available to the entire plant-based meat sector; that's a world that can start to challenge consumer expectations where product quality is concerned. And even Impossible would, I'll bet, be enthusiastic about a company that was focused on helping the entire sector bring down ingredient costs and scale up production.

Fortunately, momentum is building: More than half of all peer-reviewed articles on plant-based meat have been published since 2020.[28] And we've seen that scientific interest carry over into broader, structural signs of progress.

For example, in 2023, the world's first center focused on all alternative meats—not just cultivated meat—was announced in the nation whose government has been most enthusiastic about alternative meats: Israel. With $20 million in public funding, the Israel Institute of Technology partnered with GFI Israel to launch the Sustainable Protein Research Center.

The two countries that are vying for the "global visionary" title in alternative meats are Israel and Singapore. Entrepreneurs in Singapore

---

* I'm including in this concept all the improvements that come through "learning by doing." This includes things like improved production technologies, higher quality soy, and so on.

often tell me they have access to the entire national scientific infrastructure—from both major universities and from government scientists at the Agency for Science, Technology, and Research (A*STAR). In 2024, Enterprise Singapore, alongside funding bodies from Israel, Sweden, and Switzerland, issued a joint call for proposals focused on "plant-based, fermentation-derived and cultivated meat/seafood, hybrid products and enabling technologies . . . to promote development of alternatives to protein from living animals."[29]

That same year, the UK's Biotechnology and Biological Sciences Research Council launched a call for proposals to establish an Alternative Proteins Innovation and Knowledge Centre, backed by £15 million. Germany followed suit, allocating €38 million for alternative protein research and launching its own "proteins-of-the-future" center. Meanwhile, the University of California at Davis expanded its National Science Foundation-funded cultivated meat research to include plant-based meat, launching the Integrative Center for Alternative Meats and Proteins (iCAMP).

In 2025, the Bezos Earth Fund, Jeff Bezos's 10-billion-dollar climate philanthropy, launched three new centers of excellence at top universities selected through a rigorous process in partnership with GFI's science team. The three winners: Imperial College London, the National University of Singapore, and North Carolina State University. The fund also launched a $100 million AI initiative for climate and nature, and in its first round of submissions, alternative proteins was one of just three spotlighted focus areas and earned nine of the first 24 grants.

Also in 2025, Gordon Research Conferences (GRC) approved a proposal for a new conference on alternative meats, formally recognizing the field as a frontier area of scientific inquiry and bringing it into their prestigious series of research-focused meetings. Beginning in 2026 and every two years thereafter, a GRC focused on the science and engineering of alternative meats will assemble hundreds of brilliant minds across the global research community to talk alternative protein science with their peers.

The science of plant-based meat is trending in the right direction, but scaling is a huge concern, with no obvious solution on the horizon. Well, there is one.

The 90/10 Solution: Blends

We could solve that taste parity challenge by using commodity meat as the flavoring agent, rather than relying on expensive inputs from a handful of flavor houses. Industry experts believe that a blend of 80% plant protein or mycoprotein and 20% conventional meat could achieve taste parity with 100% animal meat at a lower cost. Sell 125 of those 80/20 burgers and you've replaced as much animal meat as 100 fully plant-based burgers.

I suspect that the best plant protein for blends will be soy, since soy protein concentrate is the least expensive of the protein options for plant-based meat by a wide margin, has the fewest off-flavors, and has the highest protein bioavailability. It also requires less land than other crops and is produced at much higher volumes. Another option is mycoprotein, which is expected to rival soy in efficiency once it reaches scale.* The cost of mycoprotein-based Quorn in the UK already points in that direction.

Getting the formulation right—hitting taste and texture parity—will still require some science. But it should be far simpler than developing a 100% plant-based product that mimics animal meat, because only a small amount of actual animal meat is needed to infuse the product with flavor. The real question is how much is needed to hit umami targets and mask any beany off-notes. Impossible has done this with an impressively small amount of soy heme and flavoring agents; soy's milder profile likely helped. How much animal meat will be required?

---

* Mycoprotein—e.g., from the companies Quorn and Better Meat Co.—is often conflated with plant-based meat, but it's actually made using biomass fermentation.

That's a great option for science, especially public access science; that said, this is also a great opportunity for entrepreneurs.

Marketing could be tricky. In a world where some products are marketed as "100% beef" and similarly, will consumers be enthusiastic about products that contain mostly plant protein or mycoprotein?

Consumer polling looks good: Our friends at NECTAR found that almost three-quarters of consumers polled in 2024 were interested or somewhat interested in blended meat. Only 4% were not at all interested. That said, blends do have a perception problem: Almost 90% of polled consumers were worried that the products would cost more or taste worse.[30]

Consumers say they prefer a higher meat content, but since we already know that 9 out of 10 consumers in the survey worried about the taste, price, or both, I'll bet that's based on a concern that a lower meat inclusion rate will lead to a less delicious or more expensive product. Once again, we'll be in a "tasting is believing" scenario, where consumer adoption will rise with familiarity.

The industry has a strong pitch if they're starting with products that compete on taste and price, since those are the most important factors to consumers. Add in health, the third most important factor, and the argument gets even stronger: Assuming we're blending with soy or mycoprotein, blends will have fewer calories (i.e., lower caloric density), less fat and saturated fat, less cholesterol, and more fiber.

Like the best-performing plant-based meats, they should also deliver a higher percentage of protein than conventional meat, and that protein will have equal bioavailability, assuming we're using soy protein or mycoprotein.

The many millions of consumers who would gladly switch to fully plant-based meat if the taste and price were right—but who still eat meat—might find blends to be the ideal entry point, until the price of fully plant-based meats comes down.

Some might worry that promoting blends will undercut momentum toward fully plant-based meats. But according to industry consultant and former GFIer Zak Weston, as long as companies are using soy protein and mycoprotein as their plant ingredient, the opposite will be true. That's because scaling up blends can help to build the protein production infrastructure that will be required for the price of 100% plant-based products to fall.

Major foodservice and retail players often complain that fully plant-based products don't sell. We know why that is: Consumers think they taste bad and cost too much. Blends can solve both problems. They offer a path to real volume and give companies a faster, more practical way to meet their sustainability and ESG goals than waiting for fully plant-based products to reach parity.

Paul Shapiro, CEO of The Better Meat Co., has been supplying plant protein to Perdue Farms since 2019. For now, Perdue's blended chicken—sold as "Chicken Plus" and available as nuggets or tenders—is a niche offering targeted at kids and sold at a price premium. But Paul tells me that his production process is efficient enough to create blends that are 80% plant protein or mycoprotein and 20% animal meat (beef, pork, or chicken) at a lower cost.

In short, blends offer a powerful near-term lever to reduce meat's environmental and public health impacts while also helping to bring fully plant-based products within reach by solving one of their biggest challenges: scale.

―――――――

For years, I believed that people just needed to understand meat's adverse impact on the global poor, nature, and animals, and they would not want to eat it. I tried, unsuccessfully, to sell my friends and family on Boca Burgers and veggie dogs. I assumed that if I liked them, others would learn to like them too.

What I failed to grasp is that for most people, food is not an ethical decision. And nowhere on earth are most people craving an almost-burger or a sort-of hot dog. They want meat. And they'll only switch if plant-based options are just as delicious and just as affordable as animal-based meat.

It took the vision of climate-driven entrepreneurs Pat Brown and Ethan Brown to convince me that meat can be made from plants: indistinguishable from animal meat in flavor and satisfaction. And because making it is so much more efficient, we should be able to scale up production and match the affordability of conventional meat.

We're on a promising trajectory. More and more science is being produced. The products are improving. Governments are starting to recognize the slew of public benefits mainstream adoption would bring.

Steep hurdles remain: Just a few products are close to taste parity with conventional animal meat, and they all cost at least twice as much. Scaling remains a beast even for ground formats.

What we need is more private sector innovation and a lot more government partnership with industry. One issue is perception: Many people simply can't fathom plant-based meat that is as delicious and affordable as animal-based meat.

The next chapter will unpack and challenge that misperception.

PART THREE

From Moonshot to Mainstream

Alt Meats for All!

CHAPTER EIGHT

The Innovation Playbook

From Inconceivable to Ubiquitous

A lot of myths about computers were exposed in 1984. One of them is that there is such a thing as the home computer market. It doesn't exist. People use computers in the home, of course, but for education and running a small business. There are not uses in the home itself.

— John Sculley, CEO, Apple, 1983–1987[1]

Our job is to figure out what [customers] want before they do. I think Henry Ford once said, 'If I'd asked customers what they wanted, they would have told me, "A faster horse"!' People don't know what they want until you show it to them.

— Steve Jobs[2]

Failures of the Human Imagination:
"It hasn't worked yet. Ergo, it can't work."

It would not surprise me if you're still dubious about the idea that your Uncle Ted is ever going to choose a big, juicy plant-based or cultivated burger over a good, old-fashioned cow-based burger. Burgers have always come from cows, you may be thinking. They always will. That's nature.

The human mind just isn't very good at imagining something happening if it has never happened before. Plants have never tasted exactly like meat; that must mean they can't taste exactly like meat. No one has ever produced actual animal meat in a factory; that must mean it can't be done.

Usually this doubt isn't fully conscious. It just feels true. When we're asked if we'd switch to plant-based burgers that were indistinguishable from animal meat, we reflexively say no. Because of course a plant-based burger cannot be indistinguishable from animal meat.

If that's your position, please take a moment to ask yourself: What's the scientific barrier that causes you to think plants can't perfectly replicate meat? Is it truly impossible, or just very hard? And if it's just hard, how hard? Have you tried the Impossible burger, Impossible nuggets, or Impossible hot dog? Most people couldn't tell the difference, and even for those who could, the products are close. Do you really think it's impossible that science can get us to parity? Why, precisely?

Now cultivated meat: What hurdle do you imagine cultivated meat can't overcome? Does Tufts biomedical engineering professor David Kaplan's confidence in cultivated meat give you pause? How about the thousands of other researchers who are dedicating their careers to the field? Or the governments funding the science? Or the meat companies investing in it? What *exactly* feels impossible to you?

Think back to when you first learned about ChatGPT. You probably hadn't heard of or imagined large language models (LLMs) like ChatGPT before they arrived. Even experts and superforecasters (people who have consistently outperformed relevant experts in predicting complex and uncertain events) who believed that ChatGPT-level capabilities were possible thought it would take decades before they reached the capacities they were showing by 2023.[3] And then almost overnight we also had Claude, Grok, Gemini, and DeepSeek.

Or think further back, before you bought your first smartphone. Until 2007, it likely hadn't occurred to you that you'd carry your camera, your music, your phone, and your email all on one device, or that you'd use that same device to hail a taxi, listen to books, find out the name of some random song that's playing at Starbucks, communicate with someone who speaks a different language, rent a bicycle, and figure out how to bypass a traffic jam on your way to the airport.

Called "convergence" in tech circles, the idea that many features would converge into one device was controversial, with many experts convinced that the idea was folly. A typical headline from a tech industry consultant and futurist reads, "Device divergence always beats convergence."[4] Less than a year later, Steve Jobs introduced the world to the iPhone.

There are endless examples of technological innovations that went from inconceivable to indispensable (or at least ubiquitous) in a matter of a few decades.

My favorite is the automobile, which emerged out of nowhere to solve one of the ugliest problems in the world's great cities at the turn of the twentieth century.

When Horse Shit Ruled the World

In 1900, Manhattan was home to around 130,000 horses. London had even more. Chicago had 75,000. Philadelphia, 51,000. St. Louis, 32,000. And your average working horse produced 30 pounds of manure and a gallon of urine a day. That meant nearly four million pounds of excrement and 130,000 gallons of urine per day on the streets of New York City alone.[5]

You might've heard one or both of these stories: The first involves an 1898 urban planning conference in New York City. The only agenda item? What to do with all the horse shit. They couldn't figure it out. The meeting was abandoned within days; everyone went home, despondent. The second story involves a prediction from the *London Times* in 1894 "that in 50 years' time, every street in London would be buried under nine feet of manure."[6]

Both stories have been widely reported in reputable outlets, and although neither story appears to be true, the filth, disease, and chaos were real, and it was as disgusting as it sounds: The streets reeked. Flies swarmed. Rain turned everything into a river of sewage. The flies carried typhoid and other diseases. Horses routinely dropped dead in the street. Owners just walked away; city workers cut the carcasses into pieces and carted them off, sometimes 15,000 in a year in New York City alone. 9,000 in Chicago. It was revolting. And it was dangerous.[7]

Solution: The Car

Inventors had been tinkering with gas-powered cars since the 1860s. The first was sold in Germany around 1890. But in the early days, cars were a novelty, and almost no one believed they'd ever replace horses.[8] When cars started showing up in cities circa 1900, they were mocked. They were hard to start, prone to breakdowns, and utterly impractical. There were no gas stations. No mechanics.

And the cost? A horse-drawn carriage might run you $25. A car cost at least $2,000.[9] Early adopters were affluent men who had the strength to crank-start the engine and the know-how to fix the vehicle when it failed. Sound familiar? It's a lot like early computers: expensive, finicky, and really just for hobbyists. One of the world's first car companies—Winton Motor Carriage Company—was led by Alexander Winton, who explained that in 1900, there was a clear consensus: Cars were too loud, unreliable, and costly to replace horses. Anyone could operate a horse.

Enter Henry Ford. At the time, he was chief engineer at Edison Illuminating Company. He began working on "horseless carriages" in the early 1890s. In 1896, he built a prototype called the Quadricycle. In 1903, he founded Ford Motor Company.* Five years later, he launched the Model T, the first car designed to be affordable and easy to drive.

Ford's innovations were simple but powerful: a steering wheel on the left for better traffic visibility, a merged engine block and crankcase for fewer leaks and lower cost, and easier shifting for a smoother, more accessible ride. The Model T wasn't just cheaper, it was better. As often happens with new innovations, the price fell over time: A decade after launch, the Model T sold for a third of its original price.

In 1900, only 8,000 cars had been built and sold in the United States. Just 13 years later, there were more cars than horses on the streets of New York City: north of 1.2 million. By the mid-1920s, the auto industry was the country's dominant industry, having produced more than 20 million cars.[10] The rest of the world soon followed, and by the late 1920s, the developed world was awash in gas-powered cars.[11]

In 1906, Woodrow Wilson, then president of Princeton University, declared that "nothing has spread socialistic feeling in this country more than the use of the automobile, a picture of the arrogance of wealth."[12] In 1906, that was probably true. And yet the technology improved quickly, becoming more user-friendly and cheaper.

The lesson? Technology can go from unimaginable to indispensable very, very quickly. For 5,000 years, people rode horses. In less than two decades, the ridiculous and unimaginable became not just unremarkable but ubiquitous.

Meat, similarly, has always come from live animals. It's easy to assume it always will.

---

* This was his third attempt; his first two car companies failed.

The Solar Revolution That Experts Missed

Solar energy and electric vehicles are also powerful examples: In 2016, solar provided just 0.5% of global energy supply. The International Energy Agency (IEA) forecast slow, *steady* growth: about 50 gigawatts per year through 2040.[13] Their reasoning? Solar was too expensive to grow more quickly. The *Economist* captured the conventional wisdom, opining that "solar power is by far the most expensive way of reducing carbon emissions."[14]

The experts were spectacularly wrong: In 2024, solar additions were 10 times the IEA's predictions for 2040.[15] The world shot past the IEA's most optimistic nine-year projection in six and was on track to beat its 24-year forecast in ten.*

By 2024, *The Economist* had reversed course: A cover story declared the "Dawn of the Solar Age." The lead article explained that a ten-fold increase in solar energy supplies would do more to reshape global electricity than building eight times the world's current fleet of nuclear reactors.[16]

From "never—too expensive!" to "inevitable!" in 10 short years.

The experts were even less optimistic about electric vehicles (EVs): In 2015, the IEA's 25-year projections for EV adoption anticipated no EV traction at all in the United States or China, and only minor in-roads in Europe, with no impact on climate change.

Why so pessimistic? The agency cited "the cost of batteries, consumer caution, and very limited recharging infrastructure."[17] *Washington Post* Deputy Opinion Editor Charles Lane voiced the conventional wisdom when he mocked EVs as absurdly expensive, lacking reasonable battery range, and—in short—meeting "few, if any, of real consumers' needs." He called any kind of EV market penetration "a delusion."[18]

---

* "Electricity Generation From Solar Power," Our World in Data, last updated May 12, 2025, https://ourworldindata.org/grapher/solar-energy-consumption?tab=chart. In 2024, the world hit 2,200 gigawatts of solar power, adding 555 gigawatts in 2024 alone. Assuming the trend since 2016 of capacity doubling every three years continues, the world will shoot past 3,200 gigawatts in 2026, fourteen years ahead of the IEA's most optimistic projections for 2040.

Fast forward: In 2016, EVs cracked 1% of US and global car sales. Eight years later, sales were up by ten times in the United States and 22 times globally.* That's 1,000% and 2,200% in eight years. Even short-term projections were crushed: The IEA's 2021 EV forecast for 2030 was surpassed just two years later.[19]

And in China, the country that scales better than any other in human history, EV sales grew from 1% to 50% in 10 years: i.e., by 5,000%.[20]

Momentum Builds: From Prediction-Busting to No Turning Back

By now, many clean energy experts believe there's no going back. In early 2025, Oxford professors Eric Beinhocker and J. Doyne Farmer wrote in *The Wall Street Journal*: "The clean energy revolution is being driven by fundamental technological and economic forces that are too strong to stop."[21]

The authors predict that solar will cost a tenth of its 2025 price in 2050, even as fossil fuel costs hold steady. Solar energy is 10,000 times cheaper than it was when it was first commercialized for the US space program in 1958, where fossil fuel costs haven't budged much over 140 years. That's because the latter are extraction-based.[22] This means that even as solar and wind costs continue to fall, fossil fuel costs won't. As this happens, renewables adoption should continue to shoot up the S-curve, eventually providing most or all electricity, unless some other form of energy—e.g., fusion—comes along and does even better.

EVs tell a similar story: Analysts from Bloomberg New Energy Finance, the IEA, and Goldman Sachs predict that once EV batteries drop to 100 dollars per kilowatt-hour, EVs will outcompete gas-powered cars

---

* "Global EV Data Explorer," International Energy Agency (IEA), May 14, 2025. https://www. iea.org/data-and-statistics/data-tools/global-ev-data-explorer (2024: 22% for the world; US: 10%).

on cost alone. Chinese battery manufacturers have already cracked that threshold.[23] When that happens globally, the transition will speed up. Goldman forecasts 50% EV adoption in the United States and 68% in the EU by 2030.[24] The path to 100% in developed economies and China seems clear.

And the shift isn't just in blue states or green economies. In 2024, Texas and Florida were the top two states for large-scale solar, and California and Florida were the top two for rooftop installations.

Two Critical Laws of Technology

Why do experts so often get it wrong?

Part of the answer is simple: Basically, human beings are just really, really bad at imagining realities that are different from the one we're living in. *Homo sapiens* have been around for 300,000 years. For nearly all that time, daily life changed slowly if at all. Agriculture is just 12,000 years old. Cities? 7,500. Rapid change—the kind that gives us cars, planes, smartphones, and ChatGPT—is a very recent phenomenon in the annals of humanity's history.

The second reason is that even experts consistently misjudge timelines, because even experts think linearly. They look at the point in time they're in, draw on their expert knowledge of decades, and can't see how what's past won't be prelude. To some degree, they're a victim of their own expertise: If you have 1,000 units of knowledge, and all 1,000 units are rooted in the current technology and the slow trajectory of progress that led to today, how can you be expected to imagine exponential adoption?

But technology often grows exponentially. Consider Moore's Law, the 1965 prediction that computing power would double every two years. Few people believed it at first, but it turned out to be right. It's the reason you have more computing power in your pocket than NASA had to land on the moon.

Wright's Law is similar: In 1936, T. P. Wright posited that for every doubling of production, the price of a technology will fall by a fixed percentage; that fixed percentage is referred to as the "learning rate" of a technology. Wright's Law represents a happy technological dance where new technologies get cheaper as we get better at building them, "learning by doing."[25] For airplanes, the figure was 20%. Same for solar: Scientists calculated that since at least 1977, solar prices have fallen by about 20% for every doubling of installed capacity.[26]

There's another law worth talking about: "Amara's Law," named for Institute for the Future founder Roy Amara, who noted that humanity tends to "overestimate the effect of a technology in the short run and underestimate the effect in the long run." Basically, people get excited about early breakthroughs and expect near-instant mass adoption. When that doesn't happen, they give up on the idea. But once a few key hurdles are cleared, adoption can suddenly take off and outperform even the original optimism.

This is the dynamic behind Gartner's Hype Cycle, which is basically Amara's law, drawn as a curve: A technology gets announced to wild excitement and grand expectations. It does not deliver, and expectations plummet, sending it into what's called the "trough of disillusionment." But if the underlying science is sound and the market incentives align, it eventually rebounds and climbs the "slope of enlightenment" toward mass adoption.

The Human Genome Project: Absurd, Dangerous, Impossible, Worthless

The Human Genome Project, which ran from 1990 to 2003, is one of my favorite examples of Amara's Law in action. It's also another case of a science project that many experts initially dismissed as impossible, impractical, pointless, or all three.

In 1985, top genomics scientists met and ruled out mapping the human genome, calling it infeasible, at least anytime soon. When the idea resurfaced the next year, it was widely derided as "absurd," "dangerous," "impossible," and "worthless." Stanford's longtime genetics chair David Botstein said the project was a waste of money, that it would almost certainly fail, and that even if it succeeded, it would produce no meaningful benefits. He saw it as a make-work project for a Department of Energy that needed something to do, since nuclear weapons were being decommissioned.[27]

Despite widespread opposition, the project had enough influential advocates that it was approved. A National Research Council (NRC) report argued that although the requisite technology for success did not yet exist, their "broad outlines" are sufficiently clear. The report suggested that "the required advanced DNA technologies would emerge from a focused effort that emphasizes pilot projects and technological development."[28]

Three years in, things looked bleak. Sequencing was slow, the tech had not materialized as hoped, and project head Francis Collins was despondent.[29] Many expected that the endeavor would either be abandoned entirely or, at best, take decades.[30] More than halfway through the 15-year project, in 1998, scientists had sequenced just one-twentieth of the genome.[31]

Five years later, in 2003, the project was completed, two years ahead of schedule—due to scientific breakthroughs and rapidly improving computer technology.

What cost $3 billion in 2003 cost $20,000 in 2010 and $200 in 2023. Scientists expect the cost to be near zero by the Project's 30-year anniversary.[32] [33] And while the early scientists could not have told you precisely what mapping the human genome would allow and skeptics like Stanford's Botstein thought it would have no useful applications at all, genomics is now central to the study of biology, human health, and medicine. Before the Human Genome Project, it would have taken *hundreds of years* to sequence the Covid-19 virus; in 2020, it took *a matter of days* to isolate, sequence, and share the virus's full genome with scientists around the world.

And to think that it almost didn't happen, because some of the world's top experts thought it couldn't work and, even if it did, that it would have no value.

Amara's Law

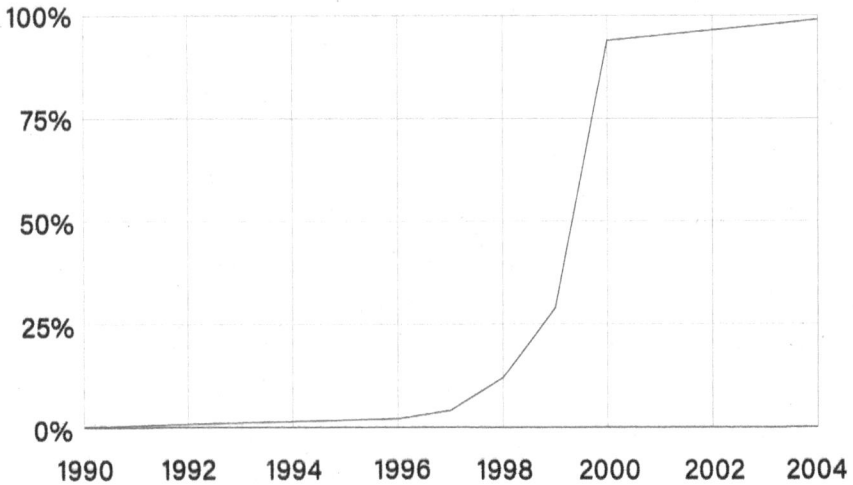

From Artificial Ice to Artificial Intelligence

Cars, solar energy, EVs, smart phones, and the Human Genome Project. There's a long history of innovations that were dismissed early, including by experts, only to become ubiquitous once key technical hurdles were cleared. What follows are a few more of my favorites.

Artificial Ice: Unnatural, Impure, Inferior

In 1806, Frederic Tudor began trying to ship ice from Massachusetts to Cuba. After years of failure and mockery, he succeeded, and by 1833, he was shipping ice to India. Dozens of companies followed, and by 1855,

Tudor had created an entire industry. By the end of the century, nearly every American home, grocer, and bar had an ice box stocked with *natural* lake ice.[34] The idea went from inconceivable (ice in India during the summer—you *must* be kidding?!) to commonplace in under 30 years.

But the natural ice industry was not long for this world.

In 1851, John Gorrie patented the first ice machine, originally for medical use. Over the next decades, "artificial ice" machines spread globally. But they weren't ideal for homes or food: too expensive, too finicky, and prone to oil or chemical contamination. Meanwhile, natural ice was cheap and familiar. The natural ice industry branded artificial ice as impure, unnatural, and inferior.

But the artificial ice industry fixed its leaks, cleaned up the product, and started running machines on electricity. That brought a major drop in cost and a huge jump in convenience. By 1914, artificial ice had surpassed natural ice in volume. Six years after that, the natural ice industry was essentially dead, hanging on only in a few rural areas. Artificial ice had taken over. Now, "artificial" ice is the only kind you'd even consider putting in your drink. It's just ice.

A few lessons seem relevant: People didn't know they wanted ice—until Frederic Tudor showed them. People didn't want "artificial ice"—until it got better and cheaper and they could see the purity benefits. Early ice innovators inspired others. The technology improved. The price dropped. And what once seemed bizarre and unnatural became normal and preferable. It's true that ice from lakes and rivers is more natural, but that doesn't make it better.

Airplanes: They Weigh More than Air, Case Closed

In 1901, the US Navy's Engineer-in-Chief called human flight "a vain fantasy." Two years later, the *New York Times* predicted it would take between one and 10 million years before humans would fly. And even that would require fixing "the existing relations between weight and strength in inorganic materials." In other words, human flight? Impossible.

In 1910, the *Guardian* dismissed the idea of airplanes in the military as useless and a waste of money: Planes flew so low to the ground that they could be shot down. Plus, planes couldn't carry more than a few passengers, making reconnaissance impossible. The editorial concluded: "We cannot understand to what practical use a flying machine that is heavier than air can be put."[35] The next year, the general who commanded the French War College, Ferdinand Foch, declared that "airplanes are interesting toys but of no military value."

Just a few years later, World War I was declared, and planes filled the skies, radically transforming military strategy.

Computers: A Dying Fad

In 1943, IBM president Thomas Watson said there might be a market for "maybe five computers." In 1977, Digital Equipment Corporation's CEO Ken Olsen declared, "There is no reason anyone would want a computer in their home."

In 1981, the IBM PC was introduced, and expectations were high. Adoption did not match expectations, and four years later, Apple cofounder Steve Wozniak called home computers "a dying fad."[36] That same year, Apple CEO John Sculley said the "home computer market" didn't exist—computers were just for education and small business.

But Steve Jobs disagreed. In a *Playboy* interview also in 1985, he explained that just as most people once couldn't imagine why anyone would want to talk to a disembodied voice on a phone, most people couldn't yet see what home computers would make possible. He predicted a nationwide communication network that would make home computers as essential as telephones. He was right of course. Most other experts were wrong.

The Dot-Com Bubble:
Shopping Online? Who Wants That?

The dot-com boom of the late 1990s is a textbook example of Amara's Law and the Gartner Hype Cycle. Between 1998 and 2000, online companies soared in value, generally without the revenue to justify those valuations. In 1999 alone, more than 100 companies doubled their value the day they listed on the stock market.

And then came the crash. The dot-com heavy NASDAQ lost nearly 80% of its value, as trillions of dollars went up in smoke. The obituaries were swift and merciless: Who would buy pet supplies online? Or trust e-commerce startups to scale? Of course the bubble popped.

But it wasn't the end. It was the beginning. In the years that followed, we got Google, Facebook, and digital streaming with Netflix. Amazon turned a modest profit in 2001 and then rocketed to $1 trillion in market value by 2018 and $2 trillion in 2024. The infrastructure caught up, consumers got broadband, and the original vision proved not just possible but transformative.

———

I could keep going: rail travel, the telephone, light bulbs, electricity, radio, washing machines, nuclear power, television, satellite communication, the internet, streaming music, and we already looked at cars, solar energy, electric vehicles, smartphones, and AI.

Most people didn't see these things coming. Early on, most people thought they wouldn't work or would never catch on. Why? Because the early prototypes were expensive and flawed.

In short, most humans assume the current state is the permanent state:

- If cars are expensive and hard to start, of course they won't replace horses.

- If ice makers are expensive and leak chemicals, of course artificial ice won't replace natural ice.
- If hooking into the internet is too cumbersome to make online shopping viable, of course no one's going to do it.
- If we've never used one device as our phone, camera, and music player, of course that won't ever happen.
- If EVs are expensive and can only drive 80 miles on a charge, of course EVs won't replace gas-powered cars.
- Sure, AI can beat Ken Jennings on Jeopardy! and Gary Kasparov at chess, but it'll never write code or pass the bar exam.

Right now, plant-based meat is expensive, and most plant-based meat does not taste much like animal-based meat. Most people believe this is plant-based meat's permanent state, if they think about plant-based meat at all. Until 2015, I believed that too. But to be clear, almost no one thinks they're making a judgment call. To say that plant-based meat doesn't taste like animal meat is like saying that the sun is hot or that water is wet. It's just true, too obvious to be worthy of further consideration or discussion.

If you're having trouble imagining plant-based and cultivated meat that are indistinguishable from industrial animal meat and that cost less, I invite you to ask yourself whether your doubts sound a little like the people who couldn't imagine cars replacing horses, artificial ice replacing natural ice, or an airplane ever flying.

Here's lesson two: Once the kinks get worked out, adoption can move far faster than most people expect, including the experts. Cars, artificial ice, airplanes, computers, the internet: Once they solved their technical flaws and brought down costs, adoption hit the S-curve, and they moved from "this will never happen" to near-universal adoption within a few decades.

I believe that plant-based and cultivated meat are the future. As discussed in the previous two chapters, there don't appear to be any scientific hurdles that would make either endeavor impossible. And since both products use a fraction of the land, water, and other inputs, first principles point to far lower costs than even conventional chicken at scale. If I'm right, then the main question is: How soon?

That's up to us; the next two chapters make the case for government partnership that could accelerate progress.

CHAPTER NINE

The Economic Case
for Alt Meats

*US leadership in alternative proteins could produce significant eco-
nomic benefits to US companies and workers, while also generating
global benefit through the shared development of novel sustainable
food technologies for a growing world.*
— The Center for Strategic and International Studies[1]

The Three-Legged Stool of Successful Innovation:
Science, Industry, and Government Cooperation

From penicillin and air travel to smart phones and the internet, most successful innovations have relied on three complementary inputs: scientific
research; government support; and industry participation. In their excellent book *Abundance*, journalists Ezra Klein and Derek Thompson use the
story of penicillin to debunk what they call "the eureka myth," the idea
that discovery or invention is the most important moment in scientific

progress. Instead, Klein and Thompson argue, much more important are the years of innovation, development, and implementation that follow.

Most readers will know the story of Alexander Fleming, the Scottish microbiologist who won the Nobel Prize in 1945 for discovering penicillin. But fewer may realize that penicillin sat on a shelf, largely unstudied, for more than a decade after Fleming's discovery. Fleming identified the compound's promise in 1928 while at St. Mary's Hospital in London, but it remained more curiosity than cure until 1939, when Oxford scientists Ernst Chain and Howard Florey figured out how to extract and purify it. They shared the Nobel with Fleming.

That's not to say Fleming gave up; he spent more than a decade mailing samples to anyone who asked, publishing papers, and advocating for penicillin's potential. But it was an unfunded side project for him, and he never managed to isolate the compound for therapeutic use or generate meaningful momentum. The degree of penicillin's neglect is captured by a 1941 article in the *British Medical Journal* claiming there had been no real interest in the drug throughout the 1930s, an error Fleming himself was forced to correct.[2]

Once Chain and Florey purified penicillin, they still needed to figure out how to scale it. In 1941, they traveled to the United States to convince US government officials that they should bankroll that final step. It worked, and an all-of-government effort was launched, led by the USDA and the newly formed Office of Scientific Research and Development. Within a few years, the US government was ordering 646 million units per month, driving the price down by 95% over the course of less than four years. Boom: Wright's Law.

So let's break that down: A publicly and philanthropically funded researcher discovers penicillin. Two more researchers, funded by the UK's Medical Research Council and the Rockefeller Foundation, figure out how to isolate it for use in humans. Then the US government steps in to scale it, funding hundreds of scientists, coordinating the effort, and then paying industry to produce the drug.

The story of penicillin's success illustrates the point Klein and Thompson were making: Discovery without development and implementation gets you nowhere. Penicillin didn't do any good at all for more than a decade after it was discovered. Plus, implementation is a lot harder and more expensive than discovery. It won't happen unless we prioritize it. Penicillin's trajectory also illustrates the importance of government cooperation with scientists and industry, since only the US government was willing to take penicillin through the final stages of development and then to production and widespread use.

In *Abundance*, Klein and Thompson lament the infinity of discoveries that have gone undiscovered. Of course we don't know what they are (since they have not been discovered), but think about these two plausible penicillin counterfactuals: First, what if the British government had invested seriously in penicillin from the moment of Fleming's discovery? The UK might have scaled penicillin first and saved countless lives during the early years of the war. But they didn't. Presumably, they didn't believe it would become useful.

Second, what if the US government had said no to Chain and Florey? What if they had seen the same amount of promise as UK officials, i.e., none? Penicillin might have sat on the shelf another decade. Or forever.[3]

Klein and Thompson point out that government support for science and innovation has been key to all or almost all scientific innovation since World War II and that a big part of the United States' position as the world's most robust economy can be traced to government support for science and engineering. This can be a hard reality for many people to accept, especially in the US, where most people are unaware of just how much government support has contributed (and still contributes) to the success of America's top industries.

In America, we're taught that progress happens mostly in the private sector and that the best thing for governments is to get out of the way. When I explain that GFI's top goal is to secure government partnership with science and industry to accelerate the progress of alternative meats, the most common concern I hear is some version of this: "Shouldn't these companies succeed or fail based on their own capacity to compete in the marketplace? Should the government really be in the better burger business?"

It's a fair question. But as I'll explore in this chapter, in today's world of global markets, successful industries achieve their success thanks to a partnership between science, industry, and government. It's very close to a tautology that industry success is a function of government cooperation with that industry; the conventional meat industry is no exception.

This observation explains why developed economies are developed: They can afford to support the industries that lead to economic growth. This is the story of the United States' booming pharmaceutical, agriculture, and tech industries, and it's the story of China's solar, EV, and manufacturing industries. It's also the story of pharma and drug production in India, the startup innovation culture in Israel, agriculture in Brazil, and so on. This kind of government partnership creates a virtuous cycle of economic growth, jobs, and tax revenues.

It's not some crazy accident that the top agricultural producers in the world are the countries whose governments pump the most money into agriculture and that the world's top hubs of scientific innovation are the countries whose governments spend the most on scientific research.[4]

The story of penicillin captures how and why this observation is true: We covered how above. But why? In short, truly new technologies are often too risky for the private sector to take them from discovery to commercialization. Success is too speculative. The timeline is too long. The research costs will be too great. No private sector actor was willing to fund the development of penicillin. The US government had to both develop the drug and then pay industry to produce it. As we'll see momentarily, penicillin is an example of the rule, not the exception.

I'm reminded of Tufts University biomedical engineering professor David Kaplan telling me that almost every week, there's some new innovation in cultivated meat that he had not even considered. All the work at Tufts, and most of the work he's talking about all over the world, is funded by governments. There's also impressive work happening across the many startups, of course, though most of that is not public. That said, as noted in chapter 6, Mark Post from Mosa Meat and Koby Nahmias from Believer Meats stand out for their many journal publications that have helped to advance the entire cultivated meat field.

If you set a Google alert for "cultivated meat," you'll be amazed at all the new ways that researchers are conceiving to bring down costs, many of them with government support.*

A Photovoltaic Revolution

The story of solar is also a story of government partnership with industry and science.† It began in 1954, when Bell Labs unveiled the first solar cell at the National Academy of Sciences and launched an ad campaign touting its promise. A front-page story in *The New York Times* called it "the beginning of one of mankind's most cherished dreams—the harnessing of the almost limitless energy of the sun." That dream stalled.

For decades, the private sector saw solar as useful only in remote loca-tions—e.g., space, research expeditions, oil exploration—and not viable at scale. That changed in the 1970s, when the OPEC oil embargo prompted President Richard Nixon to declare solar development a national priority.[5] Germany and Japan made a similar decision, driven by their own desire for energy independence. Then, in the 2000s, China made dominating solar production a top government priority.[6]

---

* You should include "cultured meat" in your alert, since cultured meat is a common nomenclature for cultivated meat.

† Greg Nemet, *How Solar Energy Became Cheap: A Model for Low-Carbon Innovation* (Routledge, 2019), https://doi.org/10.4324/9780367136604. This goes deep on this story, and it's fascinating.

Thanks to public investment across those four countries, solar prices have dropped at a consistent rate of about 20% for every doubling of installation, which has worked out to a roughly 12% price reduction annually since the Bell Labs discovery of 1954.[7] For example, between 1998 and 2023, solar prices fell from more than $7 per watt to about $0.31. In response to falling prices, solar's share of electricity grew by a factor of 35 in the 20 years leading up to 2024.[8] Wright's Law again!

Of course, this all proved quite lucrative for the solar industry, but none of it would have happened without government support for scientific research and industry scaling. Public research drove basic science. Government incentives made it worthwhile for companies to stick with solar even when profits were thin. And in countries like the United States, Germany, and Japan, the steady stream of publications coming out of government research labs created the continuity necessary for scientific discovery to proceed iteratively, even as hundreds of companies failed or abandoned the solar endeavor.*

Electrifying Transport

The history of the electric vehicle (EVs) is similar. The first electric car was introduced in 1888 by electrical engineer Gustave Trouvé, who could be the patron saint of the modern climate movement with its "electrify everything" mantra: The man's Wikipedia page lists more than 70 inventions, more than half of them powered by electricity.

Of the roughly 8,000 cars on US roads in 1900, about 1,600 were electric. But drivers wanted to go to the countryside, and there was no

---

* Nemet goes into so much more detail and is worth reading in full. He also notes that "California stepped up in the 1980s as federal budgets were being cut, and again in 2006, when solar was not a favored technology at the US federal level." Nemet, *How Solar Energy Became Cheap*, 40.

place to charge. Plus, the power grid couldn't handle an influx of EVs. So when Henry Ford perfected the assembly line, he used combustion engines.[9] The idea that EVs could ever compete was soon seen as unrealistic. Until 2016, it wasn't even enough of an idea that (almost) anyone bothered to argue against it.

There was a brief period from 1996–1999, when GM introduced the EV1, which the company marketed as a second car for city driving. But after investing about $1 billion in the project, the company pulled the plug, recalled the roughly 1,100 EVs they'd built, and crushed them. GM's perspective was that the cars were too expensive and didn't have the battery range that consumers would need to make it worth GM's while to produce them: There was not and never would be a significant market for EVs, according to the world's largest car company.

Ask someone in 2016 why EVs would never work, and they'd echo GM and the IEA's 2015 World Energy Outlook: They're too expensive, the batteries don't last, and there's nowhere to charge them. But between 2014 to 2024, battery costs fell from $715 to $115 per kilowatt-hour and battery range tripled, courtesy (in large measure) of Uncle Sam and other governments. The Department of Energy had been working on battery technology since 1976, and DOE-developed tech was central to both generations of electric cars. The breakthroughs between 2014 and 2024 required lithium-ion batteries, which were invented and developed with DOE money.[10] Finally, governments and industry worked together to build charging networks in China, the United States, and Europe.[11]

The World Bank notes that it's precisely this kind of government support that "led to massive cost reductions in key climate technologies like solar PV panels ([down] 85% between 2009 and 2019) and lithium-ion batteries used in EVs (down 89% between 2008 and 2022)."[12] The Bank recommends similar resourcing for agriculture-based technological development, including alternative proteins.

Solar Photovoltaic Module Price vs. Year

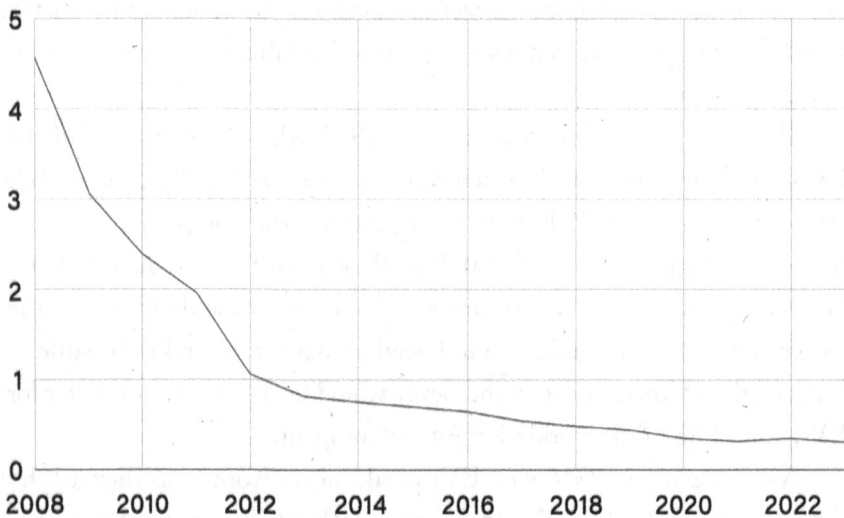

Solar Capacity (GW) vs. Year

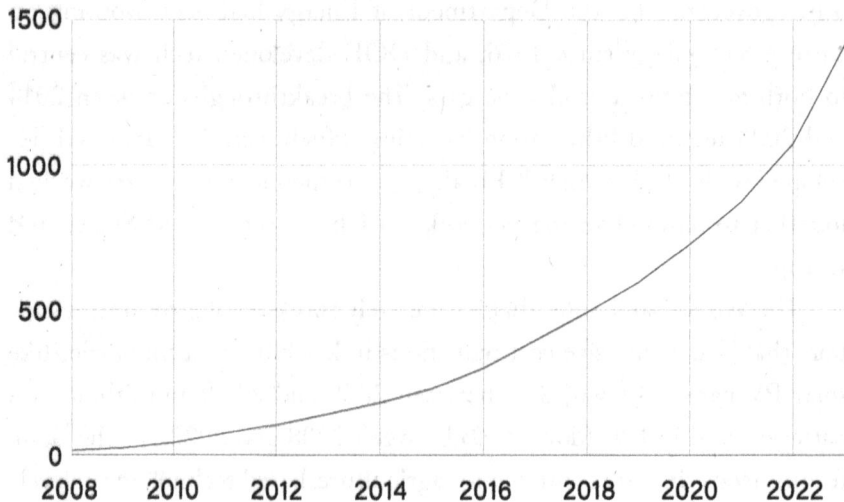

Drugs, Ag, and Tech:
Science, Government, and Industry

That kind of acceleration isn't just a win for climate, energy self-sufficiency, and cleaner air; it's a blueprint for economic growth, as becomes clear when we take a look at a few of America's most successful industries: pharmaceuticals, agriculture, and technology. I'm going to focus on the United States here, but it's worth noting that the top companies in these sectors that are not in the United States are in countries that have been supportive in similar ways.

American Pharma:
A Half-Trillion-Dollar Public-Private Partnership

The US pharmaceutical industry added $650 billion to the US economy in 2024.[13] That same year, the National Institutes of Health, which describes itself as "the nation's medical research agency," spent more than $45 billion on what it describes as "important discoveries that improve health and save lives."[14] The institute's website details the lives saved, jobs created, and economic growth created by NIH's partnership across science and industry.[15]

On its face, this might seem curious: Why is the US government spending tens of billions of dollars serving as, essentially, a research arm for private industry that brought in well north of half a trillion dollars? That's easy: If the US government doesn't partner with pharma in the United States, the industry will shift their operations to Switzerland or Denmark or the UK or another country that will.

In her brilliant exploration of industrial policy's role in economic growth, *The Entrepreneurial State*, Professor Mariana Mazzucato reports that 75% of the 357 drugs discovered in the United States between 1993 and 2004 were developed by public researchers using public funds.

Similarly, of the 356 drugs approved for sale between 2010 and 2019, NIH provided funding for all but two of them, at a cost to taxpayers of $187 billion. That's an average of more than half a billion dollars per approved drug. For the 86 most innovative drugs, those medicines that were the first to target the malady they were designed to treat, the average NIH contribution was $1.44 billion.[16]

I agree with the criticism that an industry that takes this much government largesse should not charge more for drugs in the United States than it charges for the same drugs in other developed countries, and I also agree with the critiques in both *The Entrepreneurial State* and *Abundance* that both the public and private sectors have become far too afraid of failure and too focused on short-term and incremental gains; reform is needed in both of these critical areas.

But I don't begrudge the industry its partnership with governments, because it seems clear to me that government partnership with industry is why we have the life-saving drugs that we do, as well as large pharma industries that create jobs and wealth.

Growing the Food That Feeds the World

The United States Department of Agriculture has spent at least $10 billion annually on direct payments to farmers since 1983. Direct payments clocked in at north of $15 billion annually during the eight years of the George W. Bush administration, hovered between $10 billion and $15 billion under Obama, and then shot back up during the first Trump administration. In 2020, the government transferred more than $45 billion from the federal coffers to farmers.[17] This is in addition to crop insurance payments, conservation aid, ad hoc and disaster aid, export aid, marketing aid, and taxation policies that are much more favorable for farmers than most other businesses.[18]

On top of that, public spending for agricultural research in the United States has fluctuated between 5 and 8 billion dollars per year since at least 1970. USDA describes its ag research funding as increasing "the efficiency and product quality of beef, dairy, swine, poultry, aquaculture, and sheep systems"; "developing information, tools, and technologies that can be used to improve animal production systems"; and so on.* According to USDA analysis, for every dollar spent on agricultural research, the US economy nets an additional $20 of economic activity.[19]

The European Union is even more generous: Since 2010, direct payments to farmers in Europe have hovered at 55 to 60 billion euros per year, which accounts for about one-third of total farm income, according to analysis by the European Commission.[20] According to research published in the journal *Nature Food*, more than 80% of EU subsidies support animal products.[21]

A quick aside: Many people believe that these agricultural subsidies depress meat prices and make it harder for alt meats to compete. This is almost certainly not true. In the United States, subsidies of $10 billion to $20 billion per year work out to just $0.03 to $0.06 per pound of meat and milk, and that's if all the subsidies were spent on meat and milk.† But other crops also receive substantial government support, e.g., cotton, rice, sugar, and peanuts. Plus, think about this: Governments all over the world consistently adopt protectionist measures to protect the profits of domestic meat industries (i.e., to avoid prices falling), and the meat industry has consistently been found by courts to be artificially inflating meat prices, not depressing them.

This isn't to say that agricultural subsidies are peachy keen. While the government should support farmers, I would love to see subsidies

---

* In other words, this research is focused on improving meat and dairy production.

† Using 2023 numbers from OWID for meat and milk consumption in the United States (~150 million metric tons; 331 billion pounds).

repurposed toward regenerative farming methods, as well as toward fruits, vegetables, and alt meats—more about what that might look like in the next chapter. But whatever else you want to say about subsidies, they're not a meaningful factor in the price differential between plant-based and conventional meat.

The US Economy Runs on Government-Funded Tech Research

Finally, government partnership with Big Tech: The United States is home to eight of the 10 richest humans on the planet.[22] Seven of the eight made their money in tech fields that were discovered and developed by the US government, from EVs and rocket ships to early computers, the internet, and artificial intelligence.

The federal government was also an early direct funder of many of America's formative tech startups, including Google and Apple, and it funded the computer labs at Caltech, Stanford, MIT, and other top schools for computer science, to ensure that the United States would have the most technologically advanced workforce in the world.[23] Plus, when private banks were unwilling to help, a government loan program saved the US auto industry, including Tesla.

In a particularly memorable chapter of her particularly memorable book, Prof. Mazzucato details how all 12 of the underlying technologies in Apple's most successful products—from touchscreens to the internet to email to the silicon-based chips that make all aspects of the phone possible—were developed by the federal government. Famously, when the US defense department came calling, AT&T and IBM refused to work on the internet, both because they were dubious about its likelihood of success and because it challenged their business models.[24] Since the 1950s, the federal government has been funding the foundational work in AI that made ChatGPT, Claude, and Grok possible.[25]

Sparking Jobs and GDP Growth

My thesis in this book isn't just that alternative proteins can help to solve hunger, mitigate environmental harms, and slash some key global health risks. My thesis is also that they can play a role similar to agriculture, pharma, and tech: driving economic growth and job creation.

The Center for Strategic and International Studies (CSIS) is among the most respected think tanks in the world, consistently ranking in the top five globally and often ranking first for national security.[26] In 2022, we asked CSIS to interrogate our belief that alternative proteins could benefit the US economy and address some key challenges related to food and water security. After close to two years of round tables and research, CSIS confirmed our hypothesis.[27]

CSIS reports that "US leadership in alternative proteins could produce significant economic benefits to US companies and workers [due to advantageous] trade relations between the United States and those economies expected to experience increased protein demands." The United States is already one of the world's top agricultural producers, the Center noted; alt proteins could keep the United States on top.

Those global protein markets are huge. The meat and seafood industries together bring in north of $2 trillion every year.[28] The nations that lean in on improved efficiencies for alt proteins will be the ones that command the largest share. One especially appealing opportunity for America: seafood. US production has been stagnant for decades and currently hovers at around 2% of a $500 billion global market. Seafood production globally is up 65% over the past 25 years and is expected to double by 2050. Right now, multiple of the world's most promising cultivated seafood startups are in the United States; government partnership could help them to retain their positions as the tech develops and markets grow.

CSIS also flagged biotechnology as a key global economic driver where alternative proteins could play a starring role: "The development

183

and prioritization of domestic policy that enhances economic competitiveness already exists for other strategic technologies. Investment in food biotechnology generally, and alternative protein innovation specifically, would benefit from similar treatment."

During the CSIS report launch event, Jon Bateman, who serves as co-director of the Carnegie Endowment's technology and international affairs program, noted that alternative proteins have "so many of the qualities that we're looking for: advanced manufacturing, biotechnology, high value add, IP-driven, export-led."[29] He envisioned a "virtuous cycle between alternative proteins and other aspects of the bioeconomy" and suggested that the alt protein sector could help launch new industrial hubs for twenty-first-century science and engineering.

This vision makes sense: Scientific breakthroughs in alt meat—like protein structuring, 3D printing, or media formulations—can reduce costs and improve efficiencies across other areas where these technologies are used. Already, the Australian company Vow is using cell culture to produce a cosmetics product, a flavoring agent, and another food ingredient, all of which are already cost-competitive in those market sectors. And the reverse is true too: Advances in other areas of biomanufacturing and bioprocessing could improve alternative meat production in return.*

In April 2025, a bipartisan congressional commission on biotechnology released a report calling for $15 billion in government funding for emerging biotechnologies over five years, including agricultural biotech.[30] The report urged the United States to do for biotech what it once did for pharma and tech: set up a White House office, scale public-private cooperation, reduce regulatory friction, and fund workforce development. It also recommended adding biotech mandates to a variety of key national laboratories.

---

* To a very real degree, the entire cultivated meat endeavor is based in precisely this concept: Cost-competitive cultivated meat is only possible due to advances in tissue engineering for human medicine.

From Mars to the Military

For some years, the political nonprofit Food Solutions Action (FSA)—where I'm a founding board member—has been advocating for the development of portable meat cultivators for military and space applications. FSA's military pitch centers on submarines and dangerous resupply missions. At first, this struck me as a bit far-fetched. But the congressional biotech report included a striking line, which is precisely on point: "With biotechnology, platoons will be able to synthesize food, munitions, and therapeutics directly on the front lines . . . Biotechnology will save lives on the battlefield and prevent the need for costly or dangerous refuel or resupply missions."

Space makes clear intuitive sense: The first two government grants for cultivated meat came from NASA, which supported a small project in 1995 and a more sustained effort from 1998 to 2001.[31] In more recent years, we've seen cultivated meat initiatives from the European Space Agency and SpaceX, which partnered with Israeli company Aleph Farms to include cultivated meat on the first private mission to the International Space Station.[32] [33] SpaceX's interest is logical: Its goal is to put humans on Mars. If astronauts are going to eat meat during the journey, they'll need a highly efficient way to produce it. And as SpaceX board member Steve Jurvetson told me, there's no way we're going to be using live animals for meat on Mars.

Depoliticizing Protein

The bipartisan biotech commission was chaired by Indiana Republican Todd Young and included House Republican Stephanie Bice from my home state of Oklahoma, as well as California Democrats Ro Khanna and Alex Padilla. Its bipartisan nature shouldn't be surprising: US agriculture, the bioeconomy, and national competitiveness are all bipartisan concerns, as is documented in great detail in *The Entrepreneurial State*. Across parties and ideologies, there is an understanding that these industries are driving forces for the US economy and that these partnerships have been essential to America's economic success.

The same is true for alternative proteins: In 2024, FSA worked with members of Congress to launch the Protein Innovation Caucus. Of the first 14 members, seven were Republicans, and seven were Democrats. FSA's 2024 year-in-review included two quotes on the back, both from Republican house members.

During President Trump's first term, FDA Commissioner Scott Gottlieb and agriculture secretary Sonny Perdue worked together to make the United States only the second country in history, after Singapore, to approve the sale of cultivated meat.[34] At the time, Perdue explained, "If [the United States doesn't] facilitate the invention of these ideas, we're going to see these technologies go to places around the world that are more conducive to their development, and frankly China may be one of those."[35]

It was also Donald Trump's National Science Foundation that issued the first major grant for cultivated meat outside of Israel, the Netherlands, and Singapore: $3.5 million to the University of California, Davis, "to establish the scientific and engineering foundation for the nascent cultivated meat industry, address critical scientific and engineering bottlenecks and knowledge gaps that inhibit commercialization, and train the scientists and engineers that will build the industry."[36] I'm proud to report that GFI inspired UC Davis's cultivated meat work and participated in the grant application and the consortium that came out of it.

One key issue that's motivating our Republican supporters? Competing with China. CSIS underlined the fact that "both the Biden administration and the Trump administration have emphasized increased economic competition, particularly with China, as key to US global interests. Domestic agricultural strategies—especially those that apply to rising protein demand—will be central to establishing a competitive economic advantage across future global food markets."

They are right to be concerned. Look what happened when the United States pulled back on solar and EV support in the early 2000s: China moved in, and the country now controls 80% of solar infrastructure and more than 75% of EV battery production. Solar and EV production added more than $175 billion to the Chinese economy in 2024.[37] [38]

These investments paid additional dividends for the country too: As China's economy has grown, its energy imports have skyrocketed by 800%; solar and wind have allowed it to produce more energy in-country.[39] Similarly, it wasn't that long ago that air pollution in China was front page news around the world. The shift to EVs in all of China's major cities solved that problem.

Former defense department and National Security Council official Matt Spence brought up China at the CSIS launch event as well. He emphasized that while it's essential for the US government to support existing industries like pharmaceuticals and agriculture, it's just as important to focus on "America's global competition for the next prize"—the next big innovations. In Matt's view, alternative proteins qualify.

"Food technology and innovation is happening, whether we want to be part of it or not," he explained. "When President Xi talks about investing in an innovative food solution, we know what the game plan is. So we know it's going to be happening. So now is really the time. If we want to continue to compete by having a technological lead over the rest of the world, we need to invest in these technologies now, so we're not playing catch-up like we've done with chips, 5G, and semiconductors."[40]

China has thrown down the alt proteins gauntlet, CSIS explains, "to both lessen its import burden and strengthen future food export potential." CSIS argues persuasively that the United States should compete for these global markets, noting that the United States is home to both the world's top plant-based meat companies, the world's most successful cultivated meat company, and both of the world's most successful cultivated seafood companies. This is an industry that is America's to lose.

The CSIS report also weighed in on China, both as a growing market for alt meats and as a strategic competitor. "To cede American leadership," they wrote, "is to forfeit the food security of the United States and its allies."[41]

A year later, Republican members of Congress Andrew Garbarino, Dan Newhouse, and nine others wrote to the directors of National Intelligence and Homeland Security, warning that China's aggressive investment in alternative proteins "could fundamentally alter food supply dynamics worldwide." They flagged "China's efforts to dominate emerging fields like innovative proteins, which includes cultivated meats." They also cited the DNI's annual threat assessment, which "identified China as a formidable competitor aspiring to lead the broader biotechnological landscape."

That threat, they noted, "has only been compounded by recent Chinese activity in the innovative protein market, most notably by the inclusion of cultivated meat research and development in China's five-year agricultural plan." The letter concluded: "We seek your recommendations on strategic measures the United States should consider to ensure continued leadership and resilience in this critical sector."

Alt Meats: Meat Industry Support

Another bipartisan element of alt meats is the strong meat and food industry support for these technologies. In my 2019 TED Talk, I explained GFI's stance toward JBS, Tyson, Cargill, and the rest of the meat industry: "We *need* the present meat industry. We don't want to disrupt [it]; we want to transform it. We need their economies of scale, their global supply chain, their marketing expertise, and their massive consumer base."[42]

Unlike fossil fuel companies, which have long fought energy transition, the world's largest meat and food companies are actively investing in alternative proteins. In 2016, when GFI had just eight full-time team members, I sat on a panel at Future Food Tech alongside Mary Kay James, who had just been hired to lead Tyson Foods' new venture fund. To my surprise, we expressed a similar vision: enthusiasm for plant-based and cultivated meat, a belief in consumer choice, and a shared focus on improving the taste and bringing down prices for alt meats.

In September 2022, even as Beyond Meat's stock price faltered and alt meat investor enthusiasm waned, Tyson CFO John Randall Tyson told a room full of climate policymakers that the world's largest chicken company had "made a lot of great investments in both plant-based and [cultivated meat], which we're excited about. We see that as the next frontier."[43]

No one at GFI was surprised by John's statement, since GFI had been working with Tyson for years. I had personally taken John for Impossible burgers at Jardinière in San Francisco, a few weeks before the official launch, and I also introduced him to Upside Foods' CEO Uma Valeti.

Tyson isn't alone. The world's three largest meat companies and two largest food companies are investing in plant-based and cultivated meat, and they've expressed consistent enthusiasm. We're still early in the cultivated meat journey, and plant-based products have yet to reach taste or price parity, so it's especially encouraging to see Cargill and Tyson, America's two largest meat companies, investing in multiple plant-based and cultivated meat startups.

Perhaps even more impressive, JBS, the world's largest meat company, has invested $100 million into building a cultivated meat division in Brazil, and in 2025, they bought one of Europe's most innovative plant-based meat companies, the Vegetarian Butcher.[44]

Many in the climate policy community are dubious; they've seen fossil fuel interests block clean energy and the auto industry fight public transit. It feels impossible that a major industry would support the thing that might one day replace it. But I believe two things are true: First, alternative proteins are a business opportunity for traditional meat companies in a way that renewables are not for fossil fuel companies. And second, all industries are composed of human beings, and almost every human is the hero of their own story. Human beings, when given two options of roughly equal economic value, will choose the one that makes the world better.

Let's start with the business case. Fossil fuel companies have massive capital sunk into extraction infrastructure: wells, rigs, mines, pipelines, refineries. These are expensive, long-lived assets that depend on years of production to deliver a return. Renewables threaten those investments. But meat companies don't own the feed crops, the feed mills, the farms, or the animals. The only infrastructure they typically own—slaughterhouses and processing plants—are relatively inexpensive and adaptable. They can be swapped for cultivated meat or plant-based meat facilities without much financial concern.

And animals, unlike oil reserves or coal mines, are not long-term assets. Remember: Chickens live six weeks. Pigs and turkeys, six months. Cattle and farmed salmon, two years. Meat companies aren't betting on a decades-long resource. They're in the business of supplying protein as efficiently and profitably as possible. And alternative proteins—once scaled—should be much cheaper to produce. That means higher margins.

There's also the risk mitigation factor: Animal agriculture is increasingly vulnerable to supply chain disruptions, contamination scares, zoonotic disease outbreaks, and consumer backlash. Alternative proteins reduce those risks.

Finally, the industry is run by people whose fundamental goal is noble: They want to feed the world high-quality protein. If they can make a choice that improves lives, aligns to their mandate, *and* boosts profits—well, that's a very appealing proposition.

Obviously, I could be wrong about all of this, but I will just note that when plant-based products hit turbulence in 2022, not a single one of the major players jumped ship. In addition to John Randall Tyson's strong expression of support, Nestlé's CEO predicted that every animal protein would one day have a plant-based analogue. Cargill reaffirmed its belief that the products would eventually succeed, once they nailed taste, price, and nutrition.[45, 46] Indeed, the meat and food industries have—to a company—continued to defend plant-based and cultivated meat as having a bright future, something they do not have to do.

I've seen no indication that they're opposed to alt proteins and a lot of indication that they see the same opportunities that we see.

As Cargill CTO Florian Schattenmann put it: "You have to solve the trifecta. Number one, you have to have a superior taste that either perfectly mimics meat or is a positive culinary experience. Number two, you have to get the cost right — if it's more expensive than animal protein, people will try it, but they will not come back long-term. It also has to have a positive nutritional profile."[47]

That's the alt meats theory of change—straight from the CTO of America's second-largest meat company.

State-Level Cultivated Meat Bans: Much Ado About Nothing?

Sometimes when I talk about cultivated meat, someone will bring up the handful of states that have banned it. I'm mostly unconcerned. Cultivated meat companies won't be able to supply all 50 US states anytime soon anyway. Once there are multiple companies selling their products in tony restaurants in major cities all across the country, the states that banned it will—I predict—quietly repeal these laws.

Still, I was heartened by the outcry from the right in response to the laws: From the Cato Institute to *Reason* to *National Review*, conservative and libertarian commentators all had the same basic reaction: "WTF!?"[48]

The Dispatch summed up the consensus view: "Every conservative will have the same intuition about Florida's dumb law. If there are people willing to try lab-grown meat (and there assuredly are) and there are people willing to sell it to them, by what right does the government interfere in that transaction?"

They added that it was bad economic policy, noting that the United States falling behind in cultivated meat wouldn't stop China from leading—and jobs and dollars would follow.[49]

Tom Johnston, longtime editor of meat industry trade magazine *Meatingplace*, penned my favorite response: "How many more state governments will ban the manufacture and sale of cultivated meat?" he asked. "Let's hope—for the sake of the free-market-loving, animal-based meat industry—it's zero. An industry fighting a never-ending battle for consumer trust doesn't need an association with mandates whose effects fit within the definition of fascism."

He balked at treating meat companies like endangered sea turtles, called the bans "un-American," and accused politicians of using fear to protect an industry that didn't ask for and doesn't need their help.[50]

Johnston's editorial echoed a letter by Mark Dopp, CEO of the Meat Institute, which represents 95% of US beef, pork, and poultry producers. Dopp said the bans "are bad public policy that would restrict consumer choice and stifle innovation ... Decisions about what to consume or purchase should be left to the market and consumers, not dictated by legislation that hampers progress and competition."[51]

To put an exclamation point on the matter: Within days of the Florida ban taking effect, the libertarian Institute for Justice sued the state on behalf of cultivated meat pioneer Upside Foods, calling the law anti-liberty, anti-consumer, and anti-competition.

So who is behind the bans? Small-scale cattle ranchers, mostly. Many Americans would be surprised at just how easy it is to pass a law at the state level, especially a law to ban something that doesn't actually exist. Say you're a Florida or Montana cattle rancher, and you also happen to serve in the state legislature, which is exceedingly common—half of ranchers have run for local office, and 97% of ranchers vote, according to the National Cattleman's Beef Association.[52] It's a very small request to your friends and colleagues to ban a product that does not exist and that has no constituency.

I understand why they're doing it: Cattle ranching in America is hard, and it's getting harder. From 2017 to 2022, almost one in five US cattle ranches closed up shop: 118,823 ranches. This happened even as cattle revenues were up more than 15% to almost $90 billion. Just 5% of ranches are responsible for 75% of that $90 billion. The average rancher is about 60 years old and has to work multiple jobs, which can sometimes include state assemblymember.

It makes sense that ranchers are pissed, but banning something that doesn't exist feels kind of pointless. Of course, cultivated meat has nothing at all to do with agricultural consolidation and the difficulties of modern farming and ranching, seeing as it's only ever been sold in three restaurants in the United States and had never been sold in any of the states that banned it.

In the next chapter, I'll share a vision of farming and ranching that I believe alternative proteins can make possible, because they require so much less land and water. I would love nothing more than to create an alliance behind that vision.

———

Plant-based and cultivated meat are quite likely the future of meat production; most fundamentally, cultivated meat is three times as efficient as conventional chicken and even more efficient relative to farmed fish, pork, and beef. Plant-based meat has even stronger efficiency numbers.

As production scales, the inputs for plant-based and cultivated meat will come down in price, and both products will fall below the cost of conventional meat. At that point, the food safety benefits—no bacterial contamination and drug residues, in particular—will make broad adoption likely.

The country or countries that work with their alt meat industries will be the ones that reap the economic rewards of these technologies. We've seen China do that with solar and EV battery production. For alt meats, it's really a jump ball at the moment; the countries that lean in now will have an innovation advantage going forward.

That was the optimistic perspective, but there's also a pessimistic perspective: What if these technologies don't reach their potential, because the world didn't try hard enough? These are brand-new food technologies, and while they are likely to be successful—the topic of chapter 11—that outcome is not foregone. And as already discussed, time is of the essence. Every day matters.

Alt meats have tremendous potential to mitigate hunger, protect our environment, and promote global health. Plus, the companies and countries that lean in will reap significant economic benefits.

The food security benefits are even more impressive and important; and that's the subject of the next chapter.

CHAPTER TEN

The Food (And National) Security Case for Alt Meats

The rewards of investment in alternative protein research and development are clear: realizing a food system with greater ability to provide adequate nutrition for all while mitigating global threats and enhancing US strategic competitiveness.

— The Center for Strategic and International Studies[1]

Food Security Is National Security

In the summer of 2021, GFIer Audrey Gyr mentioned that her fiancé was "on fire" for GFI's mission—not for the usual climate, global health, or animal protection reasons, but because he believed it had national security implications. That fiancé, now her husband, is Matt Spence, managing director and head of venture capital at Barclays. Matt holds advanced degrees from Stanford, Oxford, and Yale Law School, and he served as special assistant to the president and deputy assistant secretary of defense for Middle East policy during the Obama administration.

Matt told me that during his time in government, he was surprised to learn that food and water security was a top national security concern for diplomats and policymakers in the Middle East and also that the Pentagon and National Security Council considered risks related to rising pressure on arable land and fresh water to be as important to America's interests as other causes of global instability and terrorism.

Matt's argument was simple: If the world is going to consume 560 million metric tons of meat and 250 million metric tons of seafood by 2050, where will we find the land and water to produce it? The inefficiencies of conventional meat production, spelled out in chapters 1 and 2, aren't just environmental problems; they're strategic vulnerabilities.

And those vulnerabilities can cascade. Rising food prices have triggered riots and unrest in countries like Haiti, Tunisia, and Mozambique, and they've contributed to major shifts in global leadership. Food insecurity can also be weaponized, as Matt explained in *Slate* about 18 months after our first conversation: "While heading up Middle East policy at the Pentagon, I saw how ISIS leveraged drought and crop failures to win the support of vulnerable populations and expand its reach."[2] This is not hypothetical.

Because alternative meats require a fraction of the land and water of conventional meat, Matt viewed them as a national security imperative. He argued that one of the most feasible, high-impact ways the US government could reduce long-term security risks would be to invest in alternative meat production. Just as energy producers gain geopolitical leverage, food producers gain strategic influence. The United States should not cede that position.

Matt believed that a national security think tank would reach the same conclusion, and he wanted GFI to commission an analysis and report. I was less sure. I don't have a background in national security, and I wasn't sure how to frame the project in a way that would appeal to think tanks. More than that, everyone says they want to break new ground, but doing so carries risk. Would a serious institution be willing to be the first to go out on this particular limb?

To my great relief, Matt offered to help. Together, we drafted a call for proposals, and Matt joined exploratory calls with multiple think tanks. A friend in the White House suggested we target CSIS, which has strong climate and food security teams and influence with both Republicans and Democrats. He also encouraged us to include biosecurity in our scope, telling me that the White House and Congress had recently received briefings on the vulnerability of the US livestock sector to biological terrorism.

Those concerns weren't new. As far back as the George W. Bush administration, RAND had been commissioned to investigate livestock vulnerabilities and concluded that even small-scale attacks on animal agriculture could have devastating economic and public health effects.[3] Former HHS Secretary Tommy Thompson famously mused, "I cannot understand for the life of me why the terrorists have not attacked our food supply . . . it's so easy to do."[4]

We ultimately chose CSIS, which surprised me by not only engaging with the idea but embracing it. CSIS proposed a deep dive that included a thorough literature review, expert interviews, and three expert roundtables focused on economic competitiveness, global health, and biosecurity. These would be followed by white papers, a policy brief, and a public launch event. The work was led by CSIS director and deputy director of global food and water security, Caitlin Welsh and Zane Swanson, as well as CSIS director of climate and energy Joseph Majkut.

In May 2023, CSIS published their findings, which urged the US government to support the development of alternative proteins: for economic growth and job creation; to strengthen food and water security; and to reduce the risk of bioterrorism.[5]

On food security, the report cited shocks to global supply chains from Covid-19 and the war in Ukraine, writing that "the fragility of global food systems has been thrown into stark relief." Those systems are only going to be more strained: "The world 30 years from now will not only be larger . . . but wealthier, more urbanized, and consequently hungrier for meat and high-protein foods."

Alternative proteins require far less land and water. Their supply chains are shorter and more resilient. They can be produced in places where conventional animal agriculture isn't possible. In CSIS's words: "The rewards of investment in alternative protein research and development are clear: realizing a food system with greater ability to provide adequate nutrition for all while mitigating global threats and enhancing US strategic competitiveness."

On bioterrorism, CSIS echoed RAND's warning from 20 years earlier, noting how vulnerable US livestock systems are to pathogen outbreaks. The introduction of just one animal pathogen could compromise vast swaths of the meat supply. By contrast, alternative proteins can't carry animal pathogens at all, and their distributed production model is inherently more secure.

Countries Leading the Way: Food Security and Science

The themes from the CSIS report apply equally to many other countries. No country wants to be dependent on other countries for food, just like no country wants to rely on others for energy. Perhaps the two most fundamental responsibilities of any government are to keep its citizens safe and to keep them fed. In a world where arable land is shrinking and climate shocks are multiplying, those priorities are increasingly intertwined.

While "food security" means something slightly different in every national context, the countries investing most heavily in alternative proteins are motivated primarily by two overlapping concerns: economic growth and food security. Climate, biodiversity, and public health benefits may be welcome, but they're not primary drivers, other than in Europe.

The rest of this chapter explores why different nations are engaging their scientific and business communities in meat innovation. Broadly, the motivations break down as follows:

- China wants to secure food self-sufficiency or at least reverse its trend of recent decades in the opposite direction.
- India wants to eliminate malnutrition and become a developed economy in time for its 2047 independence centennial.
- The United States, Brazil, and Europe want to remain agricultural powerhouses while revitalizing their agricultural sectors.
- The high-tech, food-insecure nations of Israel, Japan, Singapore, and South Korea are seeking better food self-sufficiency and food systems resilience in the face of limited farmland.

All these countries also see economic opportunity, and some are drawn to the prestige of leading on a transformative innovation with wide-reaching global benefits.

China's Protein Security Strategy

No country imports more food than China, a vulnerability its leaders take seriously. The memory of famine is fresh in the national psyche, and the government explicitly links food security to national security. Chinese scientists played a pivotal role in the Green Revolution, and food self-sufficiency remains a top strategic priority.

President Xi repeats a core message year after year: "The rice bowls of the Chinese people must always be held firmly in our own hand and filled mainly with Chinese grain."[6] To that end, China has passed some of the world's most forward-thinking policies on food waste, farmland preservation, and food safety.

China now invests more government money into agricultural research than the European Union or the United States, and it also dominates the world in high-impact peer-reviewed journal articles, agricultural patents, and agricultural patent diffusion.[7] All that work is focused on making sure that none of the country's 1.4 billion people goes to bed hungry and that when shocks come—whether from something like the war in Ukraine, another outbreak of an animal disease that kills tens or hundreds of millions of animals, or agricultural tariffs from a top agricultural trade partner—the country is ready.

While China's rice bowls may be full, they're not full of food grown in China. The country's entry into the WTO in 2001 fueled rapid economic growth and a booming middle class, which created an unprecedented surge in meat demand. As meat consumption skyrocketed, so did imports of both meat and feed crops to produce meat.[8] Between 2000 and 2020, China's food self-sufficiency dropped from 94% to 66%, with projections of continued decline.

This dependency comes with significant risk, and the country's leaders know it. China's meat system has been repeatedly rocked by animal diseases: porcine epidemic diarrhea virus in 2011, African swine fever from 2018 to 2020, multiple strains of bird flu, and a steady stream of cattle diseases.

Alternative meats offer a strategic solution, since without live animals, animal diseases vanish. Plus, alternative meats use a fraction of the land, water, and other inputs required for conventional meat. They can help China meet rising demand while regaining self-sufficiency.

China gets it: At the end of 2021, China released its five-year agricultural plan, which included cultivated meat, as well as its first-ever bioeconomy plan, which included plant-based meat. That was followed a few months later by President Xi's speech during the country's annual political conference to spell out national priorities[9]; the president leaned in on the food security imperative of protein innovation, calling out both plant-based and cultivated meat.[10] The following year, President Xi declared that science and technology should be "the wings" on which China's food security goals are accomplished.

From 2020 to 2024, eight of the world's 20 most active cultivated meat patent holders were Chinese. Most of those are publicly funded universities, suggesting government support for open-access innovation. In 2023, Chinese cultivated meat startup CellX launched its pilot cultivated meat plant, citing favorable government policies. In 2024, Chinese government officials sampled cultivated pork at another Chinese cultivated meat startup, Joes Future Foods, and China Agricultural University hosted the World AgriFood Innovation (WAFI) Conference.

At WAFI, the Chinese government launched a report focused on Chinese food policy, the subtitle of which was "building a sustainable and diversified food supply to foster agrifood systems transformation." The report was led by two Chinese universities—China Agricultural University and Zhejiang University—and was co-sponsored by two Chinese government bodies: China's National Institute for Health and Nutrition, and the Ministry of Agriculture & Rural Affairs. The report consisted of five chapters outlining China's agricultural priorities, and alternative proteins were the subject of one of the chapters and figured prominently in three others.[11]

The report argues that alternative proteins can help China to shift back in the direction of food self-sufficiency, and it also notes the benefits for climate mitigation and environmental sustainability generally. It calls for Chinese leadership on food transformation globally and significant funding focused on creating taste- and price-competitive plant-based and cultivated meat, a swift path to regulatory approval, and encouragement of private industry to lean in on working with the government to produce delicious and affordable alternative meats.

As Shanghai-based *New York Times* contributor Jacob Dreyer noted, China is targeting "the holy grail of meat: commercially viable alternatives that taste as good as the real thing," but without the massive land use needs and supply chain fragility. And with its combination of subsidies, regulatory coordination, and domestic market scale, China may just get there first.

Dreyer sees this as a good thing, hoping that China's interest will pique the interest of the United States and other governments, driving "innovation faster and in new and necessary directions."[12]

India: Alt Meats to Solve Malnutrition

While China's alt meat investments focus on food self-sufficiency and economic competitiveness, India faces a more immediate and painful challenge: malnutrition. India is home to one-third of the world's malnourished children, more than 15% of its population is undernourished, and rates of child stunting and wasting remain alarmingly high.[13] In contrast to China, which has largely eliminated malnutrition, India struggles to meet basic nutritional benchmarks.

Alternative proteins could play a major role in changing that.

The Indian government already runs the world's largest hunger relief effort, mandated by its National Food Security Act. In 2023, it spent $142 billion supplying 800 million citizens with rice, wheat, and lentils.[14] Plus, through its midday meals program, all schoolchildren are guaranteed rice and lentils.

Shifting support toward pulses (e.g., beans, lentils, and peas) and soy would improve diets and support Indian farmers. Public health authorities, including the Food & Agriculture Centre of Excellence and the government's think tank, NITI Aayog, have flagged soy's high protein content and excellent protein bioavailability as especially valuable. But an even bigger opportunity lies in processed plant proteins and mycoprotein, which contain much more protein and more bioavailable protein than unprocessed pulses or soy.[15]

Subbing in mycoprotein or processed plant protein for wheat, rice, or dal might entirely solve India's malnutrition problem, and it could also improve the diets of the tens of millions of Indians who are not malnourished but who are deficient in protein.*

A Boon for Indian Farmers—and a Global Opportunity

Alternative meats could also provide a lifeline to Indian farmers. More than half of India's working-age population works in agriculture, yet the sector accounts for just 15% of GDP, a share that's shrinking. To support farmers, the Indian government purchases rice and wheat for public feeding programs and guarantees minimum prices through the $25 billion Minimum Support Price scheme. But the program is limited in scope, and farmers say it's not enough.[16]

Shifting procurement toward pulses and soy could improve both farmer incomes and public nutrition. The even bigger opportunity lies in using India's existing farm base as a launchpad for a plant-based protein export economy. With strong procurement policies and support for food manufacturing, India could position itself as the world's most important hub for plant-based meat's main ingredient.

India is also well-positioned to play a dominant role in the cultivated meat supply chain. The country is already a global leader in affordable pharmaceutical production and vaccine exports. The country's 3,000 pharma companies and 10,000 manufacturing plants produce one-fifth of the world's generic medicines and 60% of global vaccines, supported

---

* This strategy would also align to the Food Vision 2030 recommendation that the government should fund "innovations in product formulation with plant proteins and proteins with indigenous sources towards meeting the additional protein requirements of the vulnerable population." The Indian middle class is also consuming too little protein, as well as a surprising percentage of the Indian elite.

by the largest number of FDA-approved production facilities outside the United States.[17] That massive capacity is backed by a highly skilled scientific workforce and low-cost production ecosystem.

The most expensive components of cultivated meat, media ingredients and bioprocessing infrastructure, are similar to what's used in drug production. With support from the government, what India has achieved in pharma, it can replicate in alt meat.

This vision aligns with Prime Minister Modi's top national priority: transforming India into a developed economy by the centennial of the country's independence, 2047. Under Modi, India has doubled highway construction, ramped up solar and wind installations, expanded clean water access, and dramatically improved digital infrastructure.

More critically, India has ramped up its bioeconomy, which grew from $10 billion to $165 billion in the 10 years leading up to 2024. The country is focused on doubling that by 2030, and alternative meats are explicitly in scope. Indeed, India was an early backer of alt meat science: In 2019, the country awarded a $640,000 grant for cultivated meat research at the Centre for Cellular and Molecular Biology, the fourth government grant for cultivated meat research anywhere in the world, and the first in more than a decade.[18] Since launching in 2017, GFI India has partnered with federal and state governments, and in 2024, India's Department of Biotechnology formed official working groups for plant-based and cultivated meat and pledged more than $1 billion to a national bioeconomy strategy that explicitly includes alternative proteins.[19]

If India follows through on this momentum—investing in protein processing, cultivating export markets, and leveraging its public feeding programs—it could become the backbone of a global alt meat economy, one that solves malnutrition, supports farmers, and generates the foreign direct investment that will be a boon for India's economic aspirations.

Reimagining Agriculture: A Sustainable Future for Farmers and Ecosystems

From the late 1930s through the 1970s, farming in the United States was a viable and rewarding livelihood. New Deal programs championed by Secretary of Agriculture Henry Wallace responded to the Dust Bowl and the Great Depression with a vision of agriculture that centered on healthy and regenerative ecosystems and healthy and prosperous farm families. Farmers were paid to adopt practices that restored topsoil, to let land lie fallow when overproduction was likely, and to prioritize sustainability.

That all changed when Earl Butz became secretary of agriculture in 1971, the same year *Diet for a Small Planet* was published (ironically). Butz had no patience for what he saw as outdated conservation programs. His mantra was blunt: "Get big or get out." He reoriented US agricultural policy around maximum corn and soy production, backed by price guarantees. This removed supply-and-demand dynamics from the equation and encouraged a narrow focus on output.

The results were predictable: Countless small and mid-sized farms couldn't keep up and disappeared. By 2022, the top 1% of farms brought in more than 40% of total income, and the top 5% earned more than 75%. Meanwhile, the total number of farms fell below 2 million, the median farm earned less than $60,000, and more than half of all farms operated at a loss.

I mentioned at the end of the previous chapter the effect on ranching that continues to this day. The effects are similar across the entire farm economy: Between 2017 and 2022, 1-in-20 corn farmers and 1-in-10 soy farmers lost their farms or quit, even as corn and soy revenues grew substantially. Quite simply, a growing agricultural economy hasn't translated into prosperity for most farmers, and many now work second jobs just to survive. As one fourth-generation cattle rancher testified before Congress, "The cows no longer pay for themselves—and haven't for a very long time now."[20]

It doesn't have to be this way.

With the land savings that a shift toward alternative protein production would allow, possibilities open up for a whole new way of farming that abandons the current Butz-inspired winner-take-almost-all paradigm and embraces new priorities that put ecological health and farm families at their center.

Here's what these policies could look like: First, they would focus on high-value food for domestic human consumption. Second, they would incentivize regenerative farming practices that improve soil and ecosystem health. Third, they would enlist farmers in ecosystem restoration, including practices like agroforestry that sequester carbon. Most critically, agricultural budgets would be redirected to support this transition, with an explicit focus on ensuring that farm families are able to live the lives they deserve.

At GFI, we know this kind of transformation will not be easy, but we do believe it's possible, and we believe that a shift toward alternative proteins can help to facilitate it.

GFI Europe contracted with the UK think tank Green Alliance to model an alt protein shift across 10 European countries. Because plant-based and cultivated meat require far less land than conventional meat, replacing two-thirds of their domestic animal protein consumption with alt proteins opens up a lot of opportunities. If you divvy up the freed-up land in thirds, here's what you can achieve—and all this while producing the exact same amount of food for export:

The first third of the freed-up land could be devoted to producing food for domestic consumption. This alone would make these 10 countries food self-sufficient.[21] For farmers, the benefits of growing food that directly nourishes their communities are substantial. Crops grown for human consumption will be more lucrative than feed crops, which typically generate lower profit margins and are vulnerable to fluctuations in global commodity markets and trade disputes. Plus, farmers could shift toward pulses, which require less water and fertilizer, and actively improve soil health.

And because improved soil health also means greater water retention un-derground, pulses can help farmers deal with droughts, which are occur-ring more and more frequently all over the world.[22][23] They could also grow more fruits and vegetables, better aligning with public health goals and dietary recommendations like those from the EAT-Lancet commission.*

With the second third of the freed-up land, Green Alliance found that farmers could quadruple the area currently dedicated to regenerative farming, raising the total area farmed sustainably across the 10 surveyed countries from 9% to 36%.

In other words, even as alternative proteins decrease the amount of conventional animal protein produced, they allow for significantly more regenerative farming, which—at least for meat production—requires a lot more land. There will always be a subset of consumers who prefer "real meat," whether they align with the rugged, back-to-nature ethos popu-larized by figures like Joe Rogan and Robert F. Kennedy, Jr., or the farm-to-table, sustainability-focused values championed by Alice Waters and Michael Pollan. Green Alliance's analysis shows that alt meat production can create a lot more space, even in land-constrained Europe, for farmers to cater to these consumer preferences.

Brazil in particular, with its vast agricultural production, stands to benefit greatly from the land-use transformation enabled by alternative proteins. As deforestation pressures diminish due to decreased land needs for grazing and animal feed production, the Amazon rainforest and other critical ecosystems can be protected, reversing the trend of habitat de-struction. This shift would allow the cattle industry to pivot toward regen-erative methods, ensuring that a smaller number of cattle are managed in a way that restores soil health and enhances biodiversity.

Plus, a shift toward plant-based and cultivated meat production would create opportunities to plant a greater diversity of crops. These could in-clude indigenous plants that align with Brazil's rich agricultural heritage

* I serve on the EAT advisory council.

and could serve as sustainable inputs for plant-based meat. By embracing this transition, Brazil could emerge as a global leader in sustainable agriculture, protecting its natural resources while empowering its farmers and supporting innovative food systems.

And with the final third? This is my favorite: Governments could pay farmers to rewild land and restore ecosystems. In the EU countries modeled by Green Alliance, the emissions benefits would total almost 250 million metric tons of carbon per year—that's more than twice the emissions of all air travel into or out of the EU in 2022.[24] It's also 50 million more metric tons than the IPPC thinks possible from feeding cattle methane inhibiters and all the rest of the livestock-based interventions combined, globally.

Paying farmers for ecosystem services would also save European governments €21 billion per year in avoided tech-based carbon removal costs. Although not modeled by Green Alliance, the benefits would be even greater in Brazil and the United States, where agricultural land is vast and ecological stakes are high. Restored lands could soak up even more carbon and support biodiversity on an unmatched scale.

Vision of a New Farm Reality

A world where farmland is used in a diversified manner to produce high-quality crops and meat for human consumption domestically instead of producing cheap soy and corn for farm animals on monocropped land thousands of miles away—that's also a world with higher prices for domestic farm production.[25] Add a shift in the direction of government support for agroforestry, rewilding, and other forms of ecosystem services—all with market rates that create living wages for farm families—and you have created a world in which "get big or get out" is relegated to the dustbin of history and where farm families and entire rural communities have gone from struggling to thriving.

What I've described will require a farm policy reset. This kind of radical transformation of our food system will be harder, I suspect, than anything else I'm suggesting in this book. None of it happens automatically, even if alt proteins do gain significant traction. That said, the land use benefits of shifting toward alt proteins would make this kind of shift a heck of a lot easier.

Fortunately, the work of farm system transformation has already started: In 2021, the three UN agencies that are dedicated to agriculture, global development, and the environment called for a repurposing of $470 billion of the world's $540 billion in annual agricultural subsidies, arguing that those resources were both distorting markets and harming human health and our environment.[26] The World Bank, which was partially responsible for many of the reports cited in chapters 1 and 2, is also on board.[27]

So what's the cost to our environment and global health of the current system? A few dozen researchers from roughly half a dozen countries published their assessment, and it wasn't pretty: These experts in global health and nutrition, environmental sciences, and agricultural economics, found that perverse agricultural incentives are costing the global economy almost $20 trillion per year in health and environmental costs, which the researchers are calling "the true cost of food."[28] Their prescription? A global agreement focused on creating a fully sustainable food system by 2050.

A shift in the direction of alternative meat production could be just the jolt the world needs to make this vision a reality, giving the broader food movement some breathing room, both by decreasing pressure on land and by opening some potentially powerful alliances that feel impossible right now. The current system really isn't working for anyone. The vast majority of farmers all over the world have been left behind, and even the current economic winners are facing unprecedented challenges and precarity in the form of fragile supply chains, radical shifts in weather, and unpredictable trade policy.

Plus, if the world can agree to put a price on carbon, the shift toward alternative proteins could go a long way toward paying for the rest of the changes. Many billions of dollars have already been spent on companies that are focused on pulling carbon out of the air, called carbon dioxide removal, or CDR. The goal is to create machines that can remove carbon for $100 per ton.[29] Whether or not these machines reach commercial viability, if carbon were priced at $100 per ton, a 50% shift to alternative proteins would generate between $500 and $630 billion per year in carbon credits.* That could be used to pay for most or all of the other farm system transition goals. Plus, it seems to me that avoiding emissions in the first place is preferable to spewing them out and then removing them later.

Professors Jan Dutkiewicz and Gabriel Rosenberg point out in the journal *Logic* that the land savings alone of a shift toward alternative meats could open up new possibilities for a plethora of food movement priorities.[30] And at the CSIS policy report launch event, Schmidt Futures program scientist Genevieve Croft specifically flagged that with alternative proteins, "there is an opportunity not just to do things the way that we always have but to think more about how [this shift] can benefit society *writ large*."

Genevieve suggested that one benefit of alt proteins is the ability to place production facilities in depressed communities. Alt meat production could create "workforce development opportunities for people who may not have other economic opportunities."[31] This is true in the United States, but the impact would be even greater in countries like Brazil and India, where rural wages are especially depressed and government social welfare programs are less robust.

Genevieve's sentiments reminded me of the Gates Foundation–funded work focused on table-top mycoprotein production for smallholder farmers in Africa, discussed in chapter 1. I could easily imagine that same basic concept revitalizing cities in developed economies. Few people are

---

* Based on annual emissions decreases at 50% of between 5 and 6.4 billion metric tons of CO2eq.

enthused about an industrial farm or a slaughterhouse in their town, unless they have literally no other options, but everyone would welcome a plant-based or cultivated meat factory, which looks and operates similarly to a beer brewery (and could be just as fun!).

May I present for your consideration: A global network of friendly, neighborhood meat breweries, complete with tours and tastings and other events.

Ag Powerhouses That Intend to Stay Ag Powerhouses

The United States, Brazil, and Europe have been leaning in on the science of alt meats, while also recognizing that scale will be important in the near future. The United States was the second country after Singapore to approve cultivated meat for sale, and as I write, four US companies have been approved to sell products; that's as many as everyplace else in the world combined (two in Singapore; one in Israel; one in Australia and New Zealand).

The United States has put tens of millions of dollars into the science of alt meats from the National Science Foundation as well as the departments of energy, defense, and agriculture. State governments have also jumped in, funding research at public universities in Massachusetts, California, and Illinois. On the private sector side, the United States is just behind China in terms of number of entities securing cultivated meat patents between 2020 and 2024, and it's home to the most robust ecosystem of both the plant-based and cultivated meat companies.

Europe is on board too: The EU-funded European Institute of Innovation & Technology has been funding alt meat research for years and leading global coalitions focused on promoting the technologies. The UK government has funded more than £100 million in plant-based and cultivated meat science, and the Netherlands, Germany, and Denmark have also been global leaders. The UK's Food Standards Agency is working with GFI to develop gold standard food safety rules for cultivated meat, alongside regulators from Singapore and other countries. The UK's Foreign, Commonwealth and Development Office has funded reports and spoken globally about the importance of alt meat technologies to address climate change, biodiversity loss, and malnutrition.

GFI Brazil maintains strong relationships with the country's agricultural, science, and economic ministries, in part due to our work with the country's largest food and meat companies, helping them to move in the direction of alternative meats.[32] In 2020, the second year of GFI's competitive research grant program, the second-largest number of applicants after the United States came from Brazil. This should not have surprised us: Brazil's agricultural research is second to none globally in projects related to agricultural output.[33] Brazil's science and technology ministry has included alt meats in their work on bioeconomy production chains and their foodtech research initiative, both of which have included funding for alt meats research.[34]

The Science and Security Block

Our final food security category is what I call the "science and security" (or "tech-forward, food insecure") block: countries that have world-class science and research ecosystems while also occupying a precarious food security position, because they don't have nearly enough farmland to feed their populations. These countries are Israel, Japan, Singapore, and South Korea.

Japan and South Korea are among the top five funders of scientific research globally, the top five for journal publications, and top six for science funding as a percentage of GDP.[35] On the Information Technology and Innovation Foundation's "Hamilton Index," which tracks government spending across the 10 most important technologies for national security, they rank in the top five, alongside China, the United States, and Germany.

They are also extremely food insecure. Japan has the world's second-highest agriculture trade deficit after China and is the number two meat importer globally.[36] Prorated by population, South Korea edges out Japan on both counts. Both countries also have even lower food self-sufficiency rates than China: 37% for Japan and 44% for South Korea, the lowest among the 38 OECD countries. Like all food-insecure countries, both are actively working to improve those numbers.[37]

Israel and Singapore are even smaller than Japan and Korea. They don't expect to be global manufacturing powerhouses, so they concentrate research funding on key areas of national security, and both recognize that food security is national security.[38] Singapore boasts multiple world-class universities and a government committed to funding science, often in co-operation with China. Israel invests a higher percentage of GDP in R&D than any other country, is host to multiple elite research institutions, and maintains one of the most vibrant innovation ecosystems in the world.[39]

Singapore's agricultural trade deficit was almost $15 billion in 2022, a striking number for such a small country, and it imports 100% of its beef, pork, and chicken.[40] Israel, though about 30 times larger than Singapore, would still rank 48th in size if it were a US state. Half the country is desert, and irrigation water is scarce. It imports about 90% of its food, including nearly all its chicken feed and beef.[41]

These vulnerabilities put all four countries in a precarious position: A trade war or supply chain shock could send food prices soaring and make many foods entirely unavailable. The risk is not hypothetical: In 2022, Malaysia banned chicken exports to stabilize domestic prices, which led to chicken shortages in Singapore. This may not sound like a big deal, but GFI APAC CEO Mirte Gosker told me that poached chicken and rice is to Singapore what burgers are to the United States, so the export ban put Singapore's food self-sufficiency issues into stark relief.[42]

For all four countries, alternative meats should be a critical piece of their food security strategies.

Singapore was the first country to approve the sale of cultivated meat, in December 2020, more than two years ahead of the United States, which was second. Since then, Singapore has led efforts to build global consensus around regulatory standards, collaborating with GFI APAC and Codex Alimentarius, the global food safety standards body. They understand that for alt meats to succeed, there will have to be a global innovation ecosystem, and they're working hard to create one.

Singapore also leads on religious certification: In 2024, the country's Islamic Religious Council became the first in the world to declare that cultivated meat can be halal, a precedent with major implications for the world's nearly two billion Muslims.[43] A year later, the Korean Muslim Federation followed suit.[44] And then in May 2025, the International Islamic Fiqh Academy, an influential Saudi-based institution focused on Islamic law, did too.[45]

Singapore is also pushing scientific progress. Since 2019, Singapore has committed roughly $260 million to the Singapore Food Story R&D initiative, supported collaborative research with China, and hosted global alt meat events alongside GFI APAC. At the launch of the Bezos Centre for Sustainable Protein at the National University of Singapore, President Tharman reaffirmed the nation's commitment to global alt meats leadership.[46]

The only country that rivals Singapore's commitment to alt meats is Israel. Of the first eight cultivated meat companies in the world, three were in the United States, three in Israel, one in Japan, and one in the Netherlands. Israel ranks just behind the United States for total investment in alternative proteins, at $1.4 billion. Israeli organizations hold the third-highest number of cultivated meat patents, and two of the world's top five patent holders, Aleph Farms and Believer Meats, were founded in Israel.*

---

* The others are Upside Foods (US) at number one, Mosa Meat (the Netherlands) at number three, and Joes Future Food (China) at number four.

Israel has named alternative proteins one of its top five research priorities. At its 74th Independence Day celebration at the United Nations, Israel served an all-alt-protein menu to dozens of global ambassadors. The three speakers were Israeli President Isaac Herzog, Israeli UN Ambassador Gilad Erdan, and GFI Israel CEO Nir Goldstein, each emphasizing Israel's identity as the "alt proteins nation."

Japan has supported university-based cultivated meat research since 2020, with at least 20 projects funded in 2024 alone. Researchers are active across the country, including at the University of Tokyo, Tokyo University of Agriculture, and the Institute of Advanced Biomedical Engineering and Science. In 2020, the Japan Science and Technology Agency launched a cultivated meat and 3D bioprinting project at Osaka University, in partnership with major corporations.[47]

The Japan Association for Cellular Agriculture, founded in 2020, now has nearly 60 members, including top food and life sciences companies. It has parliamentary support and inspired the creation of a dedicated parliamentary group. In 2023, Prime Minister Fumio Kishida became the second head of government, after Israel, to declare support for alternative meats, describing cultivated meat as essential to Japan's food security strategy and sustainability goals.

In South Korea, alt protein R&D has gained momentum since 2020. South Korea now ranks third globally for cultivated meat patent filings (2020–2024), and it ties with Israel and the United States for most top 20 patent-holding organizations. In 2024, Korea's scientific research agency and agricultural ministry awarded funding to multiple alt meat projects, and the country also launched a national innovation zone to support cultivated meat science and manufacturing.

That same year, South Korea's parliament passed the Food Tech Industry Promotion Act, recognizing alt meats as a driver of GDP and jobs. The Ministry of Oceans and Fisheries followed with a cultivated seafood R&D initiative. Minister Kang Do-hyung declared: "We expect to foster a new industry that combines the traditional aquatic industry with the latest biotechnology and lay the foundation to secure a leading position in the global alternative and cultivated aquatic foods market."

A Human Genome Project for Alt Meats?

I've often suggested that some government should launch a Manhattan Project–level initiative for alternative meats. When nuclear fission was discovered in Germany in 1938, scientists feared Hitler might build an atomic bomb. Within months, Albert Einstein wrote to President Roosevelt urging action. In 1942, the United States mobilized unprecedented resources, and in just three years, the Manhattan Project delivered a terrifying new technology.

Or consider a less violent example, the Moon landing: In May 1961, President John F. Kennedy announced we would put a man on the Moon before the decade was out. The technology to do so didn't exist, and many experts thought he was nuts.[48] Eight years later, it was done.

I'll dive into this more in the next chapter, but I'm convinced that the scientific and technical hurdles for plant-based and cultivated meat are nowhere near what we faced with nuclear fission, the Moon landing, or the development of solar power and electric vehicles. We already know where we're trying to go, and the road to get there is reasonably clear. For the past five years, both innovations have developed quickly, despite minuscule resources.

Any of the countries discussed in this chapter could, through support for research and scaling, get us across the taste and price finish line, at which point domestic industries could scale.

But maybe that's the wrong model. During the CSIS launch event, Matt Spence pointed out that food security is a universal concern. Plus, it's in everyone's interest to ensure a livable planet, solve hunger and malnutrition, protect the efficacy of antibiotics, and prevent another pandemic. He reminded attendees that ISIS leveraged food insecurity to gain support, proof that hunger can drive instability and even terrorism. "This is one of the areas we need to find ways to cooperate," he said, "but we need to cooperate from a position of strength."

It's a compelling case. Solar prices fell faster than anyone expected because the United States, Japan, Germany, and China all backed early partnerships across science, industry, and government. Take any one of those four out, and solar might still be nowhere (and the world would believe that cost-competitive solar was impossible; it would not occur to almost anyone that we simply hadn't worked hard enough at it). In *How Solar Energy Became Cheap*, University of Wisconsin professor Greg Nemet highlights the global nature of the solar endeavor as key to its success.[49]

There was a similar international push for electric vehicles. Researchers analyzing 34,000 peer-reviewed journal articles on EVs found that the top international collaboration pairs were: China–UK, China–Canada, US–Canada, and US–South Korea.[50] I find a lot of hope in that kind of global cooperation and in the increasing number of global conferences and joint funding calls focused on plant-based and cultivated meat.

So maybe a better analogy than the Manhattan Project or Moon landing is the Human Genome Project, which I discussed in chapter 8 as a challenge many experts disdained as pointless, impossible, or both. The Human Genome Project was global from the start, funded by the US government and carried out by the International Human Genome Sequencing Consortium, which included 20 universities from six countries: the United States, the UK, France, Germany, Japan, and China.

Here's my suggestion: an International Alt Meats Consortium. The name needs workshopping, but the concept is clear. The United States could charge the head of the National Science Foundation or the White House Office of Science and Technology Policy with solving plant-based and cultivated meat once and for all (literally). They could assemble top scientists across relevant fields and just . . . make it happen.

But it doesn't have to be the United States. Brazil could tap the head of Embrapa, their agriculture department's research arm. The UK could appoint the head of the Biotechnology and Biological Sciences Research Council. Singapore could charge the head of A*STAR. China could appoint the president of China Agricultural University or its agriculture or science and technology ministries. India could empower its Department of Biotechnology. You get the picture.

My point is this: For a fraction of what governments already spend on agriculture or pharmaceuticals, and for a tiny fraction of what they'll spend if there's another pandemic, we could fund open-access, globally collaborative R&D on plant-based and cultivated meat. We could then give it to the world.

I can't imagine a better gift for global food and water security and, by extension, the national security of nations all over the world.

CHAPTER ELEVEN

From "Alternative Meat" to Just "Meat"

The world is filled with problems we cannot solve without more invention.

— Ezra Klein and Derek Thompson, *Abundance*

Fuck me!

— Andy Jarvis, Bezos Earth Fund

From Half a Nugget to Lunch for 200— in Less than Nine Years

In March 2025, 60 people gathered at one of D.C.'s most celebrated restaurants, Oyamel Cocina Mexicana, a fixture of the city's dining scene for more than two decades. The five-course menu was classic José Andrés, featuring two standout entrees: Salmon Saku Pacifico (pacific salmon sashimi) and Albóndigas Enchipotladas (chipotle meatballs).

The twist? The meat was cultivated, Andrés had given us a significant section of the restaurant for free, and both Republican and Democratic members of Congress were in attendance. The event was hosted by Food Solutions Action PAC, which I introduced earlier.*

Watching a room full of people, including lawmakers from both sides of the aisle, digging into cultivated meatballs from Mission Barns and cultivated salmon from Wildtype, and all as if there was nothing unusual about the meal, filled me with a deep wellspring of joy.

Earlier that day, FSA had hosted a series of events for members of Congress. We served the same products, plus a few more. A Republican member from Michigan, a self-proclaimed gourmand, tried Wildtype's salmon and declared, "My taste buds are dancing!" Another member, who holds a PhD in health policy and spent decades in public health, expressed enthusiasm for the synergies between cultivated meat and human medicine, the same connections flagged by Carnegie senior fellow Jon Bateman.

I thought back to October 2016, when I tried cultivated meat for the first time. It feels like only yesterday that Upside Foods CEO Uma Valeti brought cultivated chicken from San Francisco to New York City to share with Suzy Welch and me. GFI was just seven people back then, six of whom had been on staff for about four months.

At that point, vanishingly few people had eaten real animal meat grown without a real animal, basically just a handful of scientists and the tiny teams at the eight newly formed companies. All Uma could afford to share with us was about half a chicken nugget each, and even that probably cost something like $15,000.

Flash forward *less than nine years*, and 60 people, including multiple members of the United States Congress, were sitting down to a five-course meal at one of Washington's most beloved restaurants, enjoying full entrees of cultivated meatballs and cultivated salmon.

---

* As noted above, I am a board member of FSA.

A few months earlier, Upside CEO Uma Valeti had given a talk at TED's climate conference, called TED Countdown, and all 200 attendees were served Upside chicken as a part of the day's lunch. It was certainly the largest cultivated meat meal that had ever been served. More remarkable: When the caterer ran out of mushrooms for the appetizer, they replaced the mushrooms with cultivated meat. My mind was blown.

When Can I Eat It?

The most common question we get at GFI is about when these kinds of culinary experiences will become unremarkable. While the Oyamel dinner and the day of meetings that preceded it felt like a watershed moment, it didn't have much to say about when these products will be produced at scale. Truth be told, while six companies had received regulatory approval to sell cultivated meat by June 2025, the only places it was actually being sold were a small seafood restaurant in Oregon, courtesy of Wildtype, and in Singapore, courtesy of Good Meat and Vow.[1]

Serving 60 people at a D.C. restaurant or 200 at TED's climate conference is very different from reaching the mass market around the world. The latter is what will be necessary to make a meaningful difference across the issues I outlined in the first four chapters of this book and realize the economic and food security benefits I outlined in chapters 9 and 10. That kind of broad adoption will require products that can compete on both taste and price and millions of consumers willing to buy them.

The science of taste and price parity for both plant-based and cultivated meat seems solid and only likely to get better, as discussed in chapters 6 and 7. Infrastructure scale-up will require money, but if the money is there, there are no obvious blockers; on the money front, governments are signaling the kind of enthusiasm we'll need, as discussed in the previous chapter. And consumer surveys and first principles around what consumers care about—taste, affordability, safety, nutrition—are extremely promising with regard to mass adoption.

All that said, success is not guaranteed, and given what's at stake, pace of adoption matters.

The Rumors of My Death Are Greatly Exaggerated

It's true that consumers have not yet flocked to plant-based meat; that's because most of the products are not good enough, and all of them cost too much. It's also true that investments in plant-based and cultivated meat companies have slowed quite a bit. The reasons for that include investor expectations that timelines to commercialization for cultivated meat would be quicker and misunderstanding about what's required for plant-based meat's success. All that said, alternative meats are certainly not alone; higher interest rates, inflation, and geopolitical uncertainty have chilled startup investing across all sectors other than AI.[2]

In response to plant-based meat's flagging sales and the fundraising struggles of both plant-based and cultivated meat, there have been stories in the business press claiming that the theory of change has been tried and failed. That's wrong.

The plant-based obituaries look like this: Plant-based meat was the hot new thing, but now plant-based meat companies are struggling. Sales are down. The products are too expensive, and they don't taste good enough. *Ergo, plant-based meat has been tried and failed.*

The cultivated meat obituaries look like this: There was a lot of excitement for cultivated meat between 2019 and 2021. Startups raised $3.1 billion dollars, including from Bill Gates and Richard Branson, as well as Cargill and Tyson Foods. *Where's my cultivated burger?*

These articles miss a few important points: First, all of the plant-based meats that consumers enjoy cost a lot more than their conventional counterparts. GFI ran the numbers and found that plant-based meats cost on average about twice as much as conventional meat in 2025; plant-based chicken has an even steeper cost differential.[3] It's worse than that, though; the most meat-like plant-based products cost even more. Second, the products also don't taste good enough yet. As noted, 102 out of 122 products tested by NECTAR in 2025 seriously underperformed, and there's not a single product at taste parity yet (though six came super-close).

Cultivated meat also costs too much, the science is early, and production is just beginning. And as for that $3.1 billion invested in cultivated meat? It sounds like a lot, until you realize how thinly it's spread. That total has funded more than 150 companies worldwide, covering not just R&D but also expensive equipment, rent, production tests, salaries, and more. It's also not a one-year total; it's an all-of-time total, encompassing the first investment in October 2015 and extending through June 2025.

To put that number in perspective, consider this: The average cost to bring a single drug to market between 2009 and 2018 was $1.3 billion, and that's for a pharmaceutical company that already has all the necessary equipment and infrastructure, regulatory systems in place, and so on. That's more than twice the amount raised by the best-funded cultivated meat company, Upside Foods, for all its operations going back to its first investment in 2015.[4]

A single EV-battery factory costs between $3.5 and $5.2 billion. That's more than has been spent on all aspects of cultivated meat in all of history.[5][6]

This expectation that expensive first-generation products would displace conventional meat within a few years of first prototype would be a lot like asking why cars hadn't displaced horses in 1890, why computers hadn't replaced typewriters in 1980, or why EVs hadn't replaced gas-powered cars in 2010. In each case, the products were too expensive and did not yet satisfy consumers. Innovation and scale were required. Same thing here.

99 Problems, but the Science Ain't One

It's true that some plant-based and cultivated meat companies have struggled and that others have failed outright. News articles about these struggles have sometimes suggested that one company's struggles can be seen as a proverbial coal mine canary. But again, that misses something important about innovation: Most companies fail, and they fail not because of technological challenges but for one of the other 99 reasons that running a company is hard.

During his closing remarks at the CSIS launch event, Matt Spence offered this reminder: "If you look at any national security technology . . . they take a lot of fits and starts . . . we need to realize a healthy dose of patience and a lens into history about how other technologies have grown and evolved."[7] It's not just national security tech. I have quite a few friends who are running plant-based and cultivated meat companies, and I'm not sure I've heard a line more frequently than the great Mike Tyson truism: "Everybody has a plan, until you get punched in the face."

The biggest and most obvious challenge is the lack of capital, since without capital, you're no longer a company—no matter how strong your science. My friend Michael Grunwald jokes in his excellent book about climate and agriculture, *We Are Eating the Earth,* that at the 2019 Good Food Conference, he worried that someone might write him a check to start an alt meat company while he stood in line for coffee. Just three years later, that funding pool, like much of climate tech and food startup investment, had largely dried up.

Fundraising is just one from a limitless number of challenges for someone trying to start and then run a company. Hiring excellent personnel, renting office and research space, branding and incorporation, and on and on and on. There's so much that can go wrong, even if your science is perfect. And if you're running a company that is selling products, then the headaches shoot into the stratosphere. You would not believe the stories about ingredient suppliers and co-manufacturing plants; everything that can go wrong absolutely will.

Steven Finn's warning from chapter 7 is worth repeating: There are deep research or engineering companies, and there are consumer-facing product companies. Almost no one can do both at the same time. Trying will make execution, which is already a bitch, many times harder. One good option: If your research leads to a product that could be a great consumer product, spin out a consumer-facing company. And then keep researching.

Another way to fail is to do something that looks like the theory of change but isn't. As discussed in chapter 7, many plant-based startups expected revenue without research. They assumed they could replicate the success of Impossible Foods without spending a dime on research and development. That didn't work, and most of them failed. That had nothing at all to do with the science of plant-based meat; in fact, it was validation of its necessity.

As some plant-based and cultivated meat companies have run out of money and closed, I've seen a few founders suggest that the world simply didn't want plant-based or cultivated meat. "We built it, and they didn't come," said one. But this is, not to put too fine a point on it, nonsense: As of 2025, the world does not yet have a taste- and cost-competitive plant-based or cultivated meat product. The theory of change has not been tried and failed; it has not yet been tried.

Think about the early days of the auto industry: In 1899, roughly a decade before Ford launched the Model T, there were 30 active car companies in America. Fewer than 8,000 cars had ever been built. Over the next decade, nearly 500 new auto companies launched, with about half still operating in 1908. Just two years earlier, Woodrow Wilson famously dismissed automobiles as, essentially, nothing more than playthings for the rich. Mainstream acceptance was still far off, and almost anyone who stepped back to think about it landed in Wilson's camp: No way would these expensive and fragile horseless carriages ever serve any useful purpose.

Just 20 years later, automobiles surpassed steel, oil, and textiles to become America's largest industry. By that point, the US auto market had consolidated to just 44 companies, three of which dominated. As Henry Winston noted, "more than 500 automobile companies came in and went out in those first few years." Of course, none of their failures were indictments of the car as a product.

EVs tell a similar story. Most EV car companies have gone out of business. In 1998, GM shut down its EV division after a $1 billion investment. The field was largely left for dead. Tesla reignited interest, spawning dozens of competitors, almost all of which went belly up. Tesla almost failed, too, and would have if not for a bailout by the US government. The vacuum company Dyson pledged $2.8 billion to build EVs in 2018, only to abandon the effort three years later. The corporate graveyard of companies that topped a $1 billion valuation include Nikola, Fisker Automotive, Canoo, Faraday Future, and Lordstown Motors, each of which also raised and lost more than $1 billion.[8]

When a company collapses, it's tempting to claim that the innovation itself was simply not viable. But all those failed automobile and EV companies tell a different story. In fact, running any company is hard. It's hard for everyone, of course, but it's especially tough if your company is launching a brand-new product and is trying to compete with an incumbent that most consumers do not (yet) find wanting.

At the 2025 Tufts CellAg conference, investor Steven Finn pointed to two cultivated meat startups he knew well, New Age Meats and SciFi Foods, both once celebrated for their cutting-edge science. Their failure, he explained, was unrelated to scientific challenges. In both cases, the companies had misaligned lead investors and an insufficiently robust slate of additional investors.[9] When their lead investors decided not to participate in major funding rounds, the companies didn't have another major investor who could step in.

Making the Switch: Plenty of Land, Plenty of Steel

I'm occasionally asked where we're going to find the land to grow the crops for plant-based meat, where we're going to put all the plant-based and cultivated meat factories, or whether there is enough steel to make the cultivators and extruders we'll need. First up, land and crops: Of course, we'll need far less land and fewer crops, if we're not cycling crops through animals, as discussed in chapters 1 and 2.

What about factories and steel? Let's walk through the numbers. I'll use the United States for this exercise, because it's where I have the most data. The United States is responsible for about 13% of the global meat supply, so this analysis can tell us what would be required to replace 13% of global meat production.[10]

For cultivated meat, infrastructure needs include production plants for the media and production plants for the actual meat; in both cases, you're looking at lots and lots of steel for the vats that produce the media and meat. For plant-based meat, we'll need facilities for protein processing and to produce the meat. In both cases, there's also the machinery to process the protein and make the meat.

That 13% of the global (land animal) meat trade is 48 million metric tons, or almost 10 billion animals.[11] Let's say we're producing just a bit less meat than your average American slaughterhouse: 35,000 metric tons. We'll need 1,370 of these plants, as well as all the production equipment. In case you were wondering, you'd need 10,570 of these plants to produce the entire global meat supply.

Okay, that might seem like a lot. But now let's compare that to the production infrastructure for the American meat industry: The two main production plants are the feed mills and the slaughterhouses.

There are approximately 6,300 feed mills in the United States, churning out hundreds of millions of metric tons of animal feed.[12] First, the plant ingredients (soy, corn, and wheat mostly) are put through a grinder. Then, those ingredients are combined in massive mixers with the rest of the feed ingredients, which include chicken and pig manure, as well as millions of tons of dead cattle, pigs, and chickens.[13] Then, that final mix is forced through a pellet die under heat and pressure to form solid pellets. The pellets are then cooled to harden and prevent them from crumbling during transportation.

Finally, about 1,000 slaughterhouses for mammals and another 350 for poultry, churn out an average of 37,000 metric tons of meat per year.[14] Each slaughterhouse has a substantial amount of slaughter machinery, climate control, massive freezers, holding pens, and wastewater treatment systems.

For plant-based or cultivated meat, the 1,370 plants producing 35,000 metric tons of meat will replace the 1,350 slaughterhouses that are producing 37,000 metric tons of meat. Or we could just make our plant-based and cultivated meat plants a bit bigger. We're essentially replacing like with like.

And then the plant protein facilities for plant-based meat and media facilities for cultivated meat will replace the feed mills. The thing is, the conventional system requires more than 10 times as many feed mills and more than 10 times the feed production infrastructure as the plant-based and cultivated meat, because the latter is so much more efficient, as discussed at length in chapter 1.

And because plant-based and cultivated meat production requires dramatically fewer inputs, upstream infrastructure like monocropped farmland and fertilizer plants would also no longer be necessary at nearly the same scale.

Importantly, we're not just trying to match today's meat production; we're trying to accommodate its projected growth. If meat production stays on its current trajectory, we'll need another 3.3 billion hectares of land globally, as well as all those extra feed mills and slaughterhouses. As WRI pointed out, unless we figure out how to generate colossal feed crop productivity gains, and this in a world where climate change is leading to shrinking arable land, there will be no forests left.[15] Let that sink in.

In contrast, every incremental shift toward alternative meat *frees up* land, water, and other resources. It also frees up production infrastructure. This is both a strong incentive for industry to shift (to lower its own costs and improve its profit margin), and it also means the world can produce the same amount of meat with far fewer resources when compared to the current system.

Of course, we'll still need infrastructure, and I've heard some people speculate that steel needs for cultivated meat production might be constraining. Two thoughts: First, it's incredibly common to claim that resource constraints will make something impossible. This concept has been used to cast doubt on the solar and EV trajectories for decades: Not nearly enough land for solar scaling. Not nearly enough lithium, cobalt, and rare earth metals for EVs.[16] At least so far, that's not been true, and it hasn't even slowed these technologies down.

Where there's a profit motive, there's a way.

In fact, ramping up steel production would be far easier than finding the additional land required to scale solar production or the cobalt, lithium, and rare earth elements for EV scale-up.[17] But more importantly, that's not going to be necessary.

GFI lead scientist for bioprocessing Faraz Harsini ran the numbers and found that manufacturing enough cultivators to produce the entire global meat supply would require less than one-quarter of 1% of the steel produced in 2023, or less than 8% of stainless steel.* He notes that we'd likely need something like 50% more steel to account for piping, structural supports, and the rest of the production infrastructure. This aligns to cost calculations conducted by Ark Biotech founder and CEO Yossi Quint.† The industry could almost certainly handle this demand in one

---

* That's 1 million cultivators at 50,000 liters each. If we manufacture in 200,000-liter cultivators, less steel will be required.

† Yossi Quint's numbers: bioreactor productivity of 8.2 kg per liter; production need: 542 million metric tons (2023); bioreactor capacity: 200,000 liters; number of bioreactors needed: 330,488 (or 66,098 one-million-liter bioreactors). Cost: $528.78 billion: (200,000 liters) (330,488 bioreactors) ($8/l); see also, *Bioreactor Cost Reduction Through Cross-Industry Insights* (Good Food Institute 2025).

year without breaking a sweat; spread out over the few decades or so that will almost certainly be required for scaling up alt meats, and this is really not a concern.

One final point: Multiple companies are exploring alternatives to steel, just to save money. Plus, Vow has been slashing their steel needs through innovation focused on simplifying cultivator design, so that each cultivator uses much less steel. Other companies are doing the same thing. And all this with very limited resources.[18]

Now we're back to the human tendency to believe that the current state is permanent. In short, time after time after time, innovation finds a way.

In the final section of this final chapter, I'm going to discuss why I'm enthusiastic about the path we're on and what I think the global alt meats revolution might look like in practice.

Science Rising: Governments Join the Alt Meat Revolution

When Mark Post convened the world's first conference to explore the science of cultivated meat in October 2015, there were no scientists conducting government-sponsored science on plant-based or cultivated meat and very few who were interested in doing so. I didn't save my copy of that program, but fortunately, Mark had it and sent it to me.

By my count, of the 21 science-focused presentations, 19 of them were delivered by Mark's colleagues in tissue engineering or conventional meat sciences, recruited by Mark to speculate on how their work in human therapeutics or conventional meat sciences might apply to cultivated meat production. The only two hard science presenters who were working in cultivated meat were Mark himself and University of Bath tissue engineering professor Marianne Ellis, who we met in chapter 1.

For the next two years, there were two cultivated meat conferences per year: Mark's conference in Maastricht and a conference hosted by the cellular agriculture field-building nonprofit New Harvest. There wasn't a single conference focused on the science of plant-based meat until GFI's Good Food Conference in 2018. There were also no government grants and no research institutes. All scientific progress in both plant-based and cultivated meat was being driven by a small number of startups and a small number of scientists, using their general university funds plus small grants from New Harvest for student fellows.

That matters. Company-funded science is aimed at competitive advantage. It might solve a key challenge for that company, and when companies announce breakthroughs, that also serves as a signal of what's possible. But public sector science, especially when open access, solves scientific challenges for *everyone*. Private sector momentum was a crucial factor in persuading governments to get involved, but the real breakthroughs will come faster if we build a global scientific ecosystem. That means robust public investment.

In late 2018, GFI launched its first competitive research grants program, and in February 2019, we selected 14 university-based projects to share $3 million in funding—$1 million for cultivated meat and $2 million for plant-based, per donor instruction. That million dollars for cultivated meat more than doubled the previous decade's worth of open-access cultivated meat funding. And the $2 million more than doubled all plant-based meat open-access research funding up to that time.*

Since then, GFI has offered between $3 and $5 million in grants annually, often in partnership with national governments or the EU. Meanwhile, our science team has worked closely with academic scientists to help them secure government support, often after they received their initial project funding from us. Because of our role as a science funder, we've developed strong ties with grant-making program managers at government agencies worldwide.

---

* The projects we funded that year and in the years since can be perused at gfi.org/grants.

Now there are government-funded alt protein science centers around the world; student alt protein chapters at more than 75 top universities for science and engineering; three Bezos Earth Fund innovation centers focused on the science of alt meats; and an accelerating stream of patents and peer-reviewed papers. There are multiple academic conferences annually in every developed economy, as well as in China, India, and Brazil. I covered additional examples in chapters 6, 7, and 10, and anyone interested can follow current progress at gfi.org/newsletters.

Science Generally: Breakthroughs Paving the Way

Another reason I'm optimistic: scientific progress in adjacent fields. The *New York Times*, *Wall Street Journal*, and other major outlets report regularly on breakthroughs in tissue engineering, 3D printing, and other fields relevant to alt meats—and those breakthroughs are multiplying now that artificial intelligence (AI) is being integrated into every step of R&D.

For example, Google researchers ran chemistry simulations on a quantum computer. That same power could help us replicate the umami profile of meat using functional, affordable ingredients. When I hear that Google DeepMind has discovered 380,000 materials that are stable at low temperatures, it occurs to me that it should also be able to figure out what will be required to maximize cell growth in 200,000-liter cultivators.[19]

Physicist Mario Krenn has used AI to design experiments no human had considered, reviving long-forgotten techniques to solve stubborn physics problems.[20] At the Gladstone Institutes, AI is now being used to predict cell death—potentially game-changing for maintaining viable cell populations in cultivated meat bioreactors.[21]

Microsoft's Chris Bishop says AI can "learn the language of nature," and that this capability will allow us to tackle the world's biggest problems, from drug discovery to sustainability. The science of plant-based meat is, at its core, the science of fats, proteins, and flavors. The science of cultivated meat draws heavily from biomedical engineering. And the science of alt meats overall is about solving some of the most pressing global challenges we face.[22] AI could dramatically accelerate scientific progress for alt meats.

Allen Henderson, the early Impossible Foods scientist from chapter 7, told me he's confident the toughest problems in plant-based meat are solvable. When I asked about AI's role, he turned positively giddy. "That's going to speed everything up," he effused.

Based on our work with them, the Bezos Earth Fund announced that it would include alternative proteins as one of three topic areas for their $100 million challenge for climate and nature. Of the 24 grantees in round one, nine were focused on alt proteins. Former GFI vice president for science and technology Liz Specht served as one of the three alt proteins judges, alongside AI scientists from Google and Salesforce.[23] After the BEF first-round awards were announced, Liz shared her enthusiasm that "the challenges that exist in the alt protein field are concrete enough to articulate viable AI-powered approaches to address them, something that can't be said for many fairly nascent scientific endeavors."

Another big AI upside: translation. Large language models are making science more global. Even before GFI launched affiliates in Japan and South Korea, we were hearing from researchers in both countries who were translating and adapting our scientific white papers for local use. Now, as language barriers fall, global collaboration is speeding up even more.

Private Sector Resilience

The private sector is also ignoring the death notices. As just one from among many examples, Australia-based Vow started selling cultivated quail in Singapore in April 2024. Vow was founded in 2019 by biochemist and chef George Peppou, and it took its quail from idea to commercialization in just five years, and on a total budget of less than $10 million per year. That includes *everything*: not just R&D, but also operations, staffing, and infrastructure. On top of that, they navigated Singapore's regulatory process and became the first company approved in Australia and New Zealand. That timeframe and budget are extraordinary for a product that wasn't for sale anywhere on the planet until late 2020 and still has no production infrastructure to build on.

Vow launched with a 15,000-liter cultivator capable of producing around 11 metric tons of cultivated meat per year. In 2025, they added another cultivator, this one at 22,000 liters.[24] Founder and CEO George Peppou told me that this bioreactor can produce 1,500 to 2,000 kilograms per week of meat, so that's another 78 to 104 metric tons per year.

A big part of Vow's success, George told me, has been a relentless focus on process simplicity, especially in their cultivators, where the meat is grown. That makes sense when you learn that Vow's chief technology officer, Ines Lizaur, spent five years as an engineer at SpaceX, a company famous for simplifying complex systems to cut costs, after earning her mechanical engineering degree from Stanford.

Meanwhile, the companies that supplied meat for FSA's congressional dinner, Mission Barns and Wildtype, are now producing cultivated meat and seafood on an ongoing basis, and they're also doing that on a shockingly tight budget, considering just how recent the entire cultivated meat endeavor is: Mission Barns, founded in 2018, has raised just $120 million. Wildtype, founded in 2017, has raised just $140 million, including from Leonardo DiCaprio, who shared on X: "Wild fish populations are threatened more than ever before. I'm pleased to be an investor in Wildtype . . . which will give us the chance to protect our oceans while creating the cleanest seafood on the planet."

In 2023, Wildtype finished construction on their second cultivated seafood pilot plant by upgrading a former brewery, and they have been producing salmon there ever since. I've been there repeatedly in my capacity as a board member of Food Solutions Action, as we have introduced both Senators and House members, Republican and Democrat, to the wonders of mercury-free salmon, produced in a former brewery. Enthusiasm is universal and bipartisan.

Co-founders Justin Kolbeck and Aryé Elfenbein are great ambassadors for the ålt meats endeavor: Justin received his MBA from Yale and served for six years as a foreign service officer in Australia, Pakistan, and Afghanistan. Aryé received his PhD in vascular biology and an MD, both from Dartmouth. He trained in cardiology at Yale's medical school and the Gladstone Institutes before co-founding Wildtype with Justin.

They constructed their current plant for just $15 million and are committed to what Justin calls the "scrappy, modular approach" to increasing capacity, about 250 tons at a time. This allows them to be demand-driven and to learn and upgrade as they go. They believe they'll hit price and taste parity with conventional salmon once they reach something like 1,000 to 1,500 tons of capacity, which will be a function of demand and could happen quickly.

There are many additional examples, including industry front runner Upside Foods, which launched their Berkeley, California, production plant in 2021; BlueNalu, which is producing cultivated blue-fin tuna at their pilot plant in San Diego; Good Meat, which is producing at their pilot plant in San Francisco and through a Singapore co-manufacturer; and many more, all over the world.

In the meantime, there are contract manufacturers like the Cultured Hub in Switzerland, which opened in 2024 and is the largest cultivated meat contract manufacturer in the world. And the State of Massachusetts granted $2.1 million to Tufts to establish a pilot facility on campus.

Another exciting development: The rise of companies focused on single parts of the supply chain. Today, there are cultivated meat startups specializing in media formulations, cell line development, and cultivator design. These companies exist to license their innovations broadly, meaning their breakthroughs have sector-wide benefits similar to open-access research.

As just one from among many examples, Tim Olsen earned a

bachelor's, master's, and PhD in bioengineering and then worked for years as a biomedical engineer. From 2020 through 2024, he oversaw the cultivated meat work for the German pharma company Merck KGaA, managing a global team of more than 20 Merck scientists. In 2024, he left to start a cultivated meat company focused on supplying bioreactors to the rest of the industry, with a focus on structured meat products.

As noted in chapter 7, the plant-based sector could use some similarly focused companies that are conducting research into soy and mycoprotein, into fats and flavors, and into better production machinery. Companies that are focused on solving these challenges and then licensing their discoveries to the entire plant-based meat industry could be both game-changing and incredibly profitable. There are dozens of these kinds of companies on the cultivated meat side; the fact that so few exist on the plant-based side presents a very appealing opportunity for entrepreneurship.

Okay, so for plant-based and cultivated meat, public sector and private sector science are on track. Does that mean we're on a glide path to success—it's just a matter of time before McDonald's and KFC are selling as much alt meat as they're selling conventional meat?

Probably, yes, but "just a matter of time" is doing a lot of work in that question; I'm reminded of the Steven Wright joke that "everything is walking distance if you have the time." Put another way, that glide path to success could take a (very long) while on our current trajectory.

The most oft-quoted line in cultivated meat circles comes from Winston Churchill, who predicted cultivated meat in a 1931 essay. He wrote that humanity would "escape the absurdity of growing a whole chicken in order to eat the breast or wing, by growing these parts separately under a suitable medium."[25] Although wrong on the timing—he thought it would happen by 1981—I think he's right that it will happen. But when?

In April 2021, I told *New York Times* journalist Ezra Klein that government partnership was going to be essential to the success of alternative meats: "If we leave this endeavor to the tender mercies of the market, there will be vanishingly few products to choose from—and it'll take a very long time," I predicted.[26] Four years later, we had more plant-based options that are within striking distance of taste parity than I would have predicted based on the lack of government support and dried-up investment ecosystem, and we had made far more progress on the science of cultivated meat as well. Plus, AI was opening all kinds of scientific and engineering opportunities that no one had previously thought possible.

It's a mantra of this book that we need more government support for plant-based and cultivated meat science, and when I surveyed more than 50 plant-based and cultivated meat scientific leaders in early 2025, all the plant-based scientists were enthusiastic about government support for plant-based research. On the cultivated meat side, however, there were a handful of outliers, who said the science was progressing quickly enough and that all they really needed was support to build factories.

That was the view of Believer Meats founder and chief science officer Koby Nahmias, who reminded me that "there is a history of food innovation in the United States that is as old as Harland Sanders. While the government did support the development of new breeds of chickens, it was entrepreneurs like John W. Tyson that made chicken production cost-effective." Multiple others agreed, though they also tended to grant that science funding would speed global adoption by creating more competition.

Solving the Scaling Conundrum

Scaling is the top concern for almost all the more than 50 scientists I polled, on both the plant-based and cultivated sides: A small commercial cultivated meat facility might cost between $100 and $150 million. I heard from multiple scientists working for companies that a low-interest, long-term loan at that amount could get them to profitability, either be-cause they would be able to sell their meat at premium prices (e.g., the seafood and foie gras companies) or because they could blend their meat with plant proteins to bring down cost.

But standard bank loans that size are not available without sufficient sales, and sufficient sales are impossible without a production facility. I hear the chicken-and-egg analogy from alt meat CEOs almost as much as the Mike Tyson line about everyone having a plan until they get punched in the face.

So what is the scaling solution? There are a million ways this could happen, but here are three that seem particularly viable to me: Working with governments; working with big food, meat, or pharma giants; or working with an innovation-forward tech company.

At the beginning of 2025, there were 155 cultivated meat companies operating across 34 countries. Europe leads with 53 companies across 18 countries (including 16 in the UK), followed by 36 in the United States and 18 in Israel.[27] With those numbers, it seems likely that some of them will form the partnerships necessary to fund small commercial factories.

The most obvious potential partners are governments or major corpo-rations in relevant sectors (food, meat, pharma), which could work with startups to build small or even mid-sized commercial plants. There are a range of options: Governments or major corporations could guarantee standard bank loans, governments could issue the loans themselves, or food or meat corporations could enter into manufacturing and distribu-tion agreements with cultivated meat companies, helping them to scale directly.

It may require just one or two successful facilities to prove the model and unlock a wave of new investment. With just a few more successes, momentum will build, allowing the scale-up that leads to falling prices.

Another promising option: partnership with one of the world's major tech giants. Google, Microsoft, or Amazon could guarantee a loan for scale-up and adopt plant-based or cultivated meat as an AI challenge, allocating the requisite resources, staffing the endeavor with elite scientists and engineers, and aggressively prototyping solutions, all alongside the startup and all with a heavy dose of AI learning.

Welcome to McDonald's, May I Take Your Order?

When alt meats were at their hottest, Credit Suisse predicted that they could capture between one-quarter and one-half of the global meat market by 2050, generating sales of between $555 Billion and $1.1 Trillion. Barclays predicted that cultivated meat alone could capture 20% of the market by 2040 and double that share by 2050. McKinsey echoed that mid-century 50% figure but noted it would require significant public investment in science and scaling.[28]

If we can maintain scientific momentum and unlock meaningful public and private investment, these projections feel plausible to me, albeit ambitious. With significant infusions of public sector support or a Google X or similar effort, we might even crush them, because a first principles analysis could not be stronger.

What do I mean by that? Just that the efficiencies of plant-based and cultivated meat are so much better than conventional meat, and there are no inputs into either process that should be limiting. So as production scales, prices should fall to well below the costs of conventional production. At some point, Wright's Law bumps up against resource constraints, but for plant-based and cultivated meat, that should not happen until the products are much less expensive than conventionally produced meats.

Chris Davis, who served as research and development lead at Impossible for more than a decade, confirms this thinking with regard to plant-based meat. Chris told me both that plant-based science is on the right trajectory for taste parity across the range of products and that if the companies can raise money to scale, eventual price parity is baked in. "At global scale the dominant cost is raw materials, and therefore the inherent gain in efficiency of producing meat directly from plants rather than using animals to upcycle the plants will prevail."

That said, we have a lot of work to do, and the scaling challenge for plant-based may be even harder than the scaling challenge for cultivated, because so few companies are tackling the plant-based science challenge. We still have some basic science questions we need to answer, of course, but I'm optimistic that we're on the right track to answer them. I discussed both halves of that sentence in chapter 7, so I won't rehash it here.

What about cultivated meat? No one is paying more attention to the factors that will allow us to answer this question than GFI lead scientist for cultivated meat Elliot Swartz, who told me in 2025 that the two primary cost drivers for cultivated meat, media costs and bioreactor costs, had already fallen much more quickly than we expected just four years earlier.

Media costs had fallen by more than 99% relative to the biopharma benchmark, with no sign of slowing down.[29] And cultivated meat startups reported much lower cultivator costs as well; remove all the biopharma bells and whistles, and even a custom-made cultivator costs about one-fifth what the same piece of machinery would cost for pharma.

In Elliot's estimation, right now, cultivated meat companies can compete with high-end products like bluefin tuna or foie gras. Tim Olsen said something similar: that he has "a high level of confidence that cultivated meat products can be brought to market in high-end restaurant settings at price parity" right now. That appears to be very close to a scientific consensus, and we're seeing the first steps of that strategy from Vow and Wildtype, which are selling right now in elite restaurants.

As with plant-based meat, the first principles analysis couldn't be much better, Elliot pointed out: Cultivated meat has a feed conversion ratio that is one-third that of chicken. As media formulations and production become more efficient, that should improve. "If humanity wants this to be cheap, we can make it cheap," Elliot concluded.

Alternative Meats: The Expert Consensus

When I asked the 27 cultivated meat scientists I polled in 2025 to predict how quickly the product would reach 1%, 10%, and 50% market share, pretty much everyone told me that we'd have no real idea until a handful of companies were producing in commercial plants. Once that was happening, we'd see prices fall and sales grow. Quite a few invoked the concept of S-curve growth from chapter 8.

For example, Aleph Farms founder and CEO Didier Toubia, who sent cultivated meat to the International Space Station in a joint project with SpaceX, said that cultivated meat would take the same amount of time to go from 0% to 5% as it takes to go from 5% to 50%. He guesstimated 10 years to 5% and 10 more years to 50%.

We might take a cue from what happened with cars. In 1900, only about 8,000 cars had ever been sold in the United States. By 1908—the year Ford introduced the Model T—sales had grown to nearly 200,000.[30] Ford improved functionality and brought prices down, and within five years, cars were more popular than horse-drawn carriages. Within 25 years, there were tens of millions.

Or look at EVs: Once they were approaching cost and convenience parity with gas-powered cars, adoptions shot up: from 1% to 50% in China in 10 years, and from 1% to 22% globally in eight.[31] And they're still a bit more expensive everywhere on the planet other than China, Norway, and Sweden—the three governments that have been most supportive of the EV transition.

The comparisons are not perfect, of course, but they illustrate the speed at which a shift can happen. We might even beat the early 1900s conventional-car trajectory, since we're a century more technologically sophisticated than we were in the 1910s. We might also crush the EV trajectory; consider that the auto industry is twice the size of the meat and seafood industries combined, $4 trillion per year.[32] Plus, the EV infrastructure and raw materials challenges are much, much steeper than the alt meats infrastructure challenges. Once alt meats hit price and taste parity, scaling could happen fast.

And so I'm back to the value of government partnership with industry. We're seeing encouraging government involvement all over the world, and because of the economic and food security incentives, I'm optimistic about that involvement continuing. In my estimation, the alt meats transition is on a very likely path to eventual success. But in countries where governments do not increase their support for science and scaling, there will be some painful hiccups on the journey. That will include the failure of even more companies with excellent technology.

My final point: Governments can and should ensure that as the plant-based and cultivated meat sectors grow, they maximize benefits to all. *Nature Biotechnology*'s editors noted the need for governments to support cultivated meat in tandem with policies to use the freed-up land for "biodiversity restoration, carbon sequestration, small farmers and agroecological farming." Genevieve Croft from Schmidt Futures talked at the CSIS event about ensuring benefits for rural communities. Jan Dutkiewicz and Gabriel Rosenberg presented alt meats as a chance to reset farm policy in favor of "consumers, workers, animals, and the environment."

Jan and Gabriel also talked about the democratizing capacity of government research that lowers barriers to entry for entrepreneurs, creating more competition and a much more decentralized meat industry.

Here's my vision for alt meat production: plant-based and cultivated meat production facilities taking root in rural and urban communities in every corner of the world, retrofitting and repurposing abandoned, underutilized, or legacy infrastructure, bringing new jobs and economic growth to small towns and big cities alike, everywhere people eat.

From "Alternative Meat" to Just "Meat"

Could this vision become reality by 2050? The automobile took over the world in 25 years. The electric vehicle may be on track to do the same. On the other hand, penicillin languished on a shelf for more than a decade, and it came close to languishing for even longer.

Alt meats might become just "meat" in the fairly near future—but only if we want it.

The Ripple Effect

Ushering in the Next Agricultural Revolution

It's very hard to achieve things you're not trying to do.

— Joseph Romm[1]

When Can We Drop the "Alt" from "Alt Meats"? That's Up to You

Humans have eaten meat for about 2.6 million years, and we've farmed animals for about 12,000. We now consume more than 550 million metric tons of meat and seafood, a number that has been rising since we started keeping track in 1961 and is expected to continue rising at least through 2050; that's as long as anyone's forecasting.

But now, science is offering us a better way to produce the meat that humanity craves. If we embrace it, we can cut climate emissions, preserve nature and forests, lower the price of nutritious food in developing countries, slash AMR and pandemic risk, and save countless animals from the miseries of industrial farms and slaughterhouses.

I spent January through June of 2025 emailing and talking with more than 50 brilliant plant-based and cultivated meat scientists, asking everyone whether they thought plant-based and cultivated meat could reach price and taste parity with conventionally produced chicken (I chose chicken because it's the least expensive meat; if we can solve chicken, we can solve it all). I also spoke at alt meat conferences at the University of Toronto, Tufts, and Harvard, alongside scientists from these and other universities working to solve the challenges of plant-based and cultivated meat parity. So I asked them too.

These are the scientists who are closest to the work, and they are intimately familiar with the technical and other challenges. They also see and hear all the short-term, quick-take narratives playing out in the news.

They are, to a person, incredibly optimistic about the likelihood of success for the alt meats endeavor.

But success is not self-executing. There remain steep scientific challenges. There remain even steeper scaling challenges.

Plus, the sooner we make progress, the more good we'll accomplish. All the harms of industrial meat production multiply over time. Speed matters. A lot.

The power of one person to drive meaningful change is incalculable. One individual, fully dedicated to advancing alternative proteins, can do incredibly powerful direct work. Even more inspiring: That same individual can create a ripple effect: shifting an organization, a university, even an entire profession.

That's power. That's how we win. That's how we create the next agricultural revolution.

Being part of it feels to me like a life well-lived.

What follows are some ways to plug in.

Saving the World Through Science

GFI's top goal is to build a thriving alt meats scientific ecosystem that will drive the breakthroughs we need for success. That means we need plant biologists, tissue and mechanical engineers, and every other kind of scientist. If you're a scientist, that means you. If you're just entering college, you again: Be a scientist or an engineer.

Here's an insight that I've somehow managed to neglect until now: We don't have enough engineers. Probably the toughest challenge in cultivated meat is optimizing the cultivators; no one has ever grown tissue at 50,000 liters, let alone 200,000 or a million. That's an engineering problem. Similarly, one of the toughest challenges in plant-based meat is the lack of fit-for-purpose machinery that can create the perfect texture across all kinds of plant-based meats. For both, we need far more people working on more efficient, low-cost, scalable production technologies. Help!

The coolest thing about university, if you ask me: the opportunity for massively outsized impact. One student can convince tissue engineering professors to include cultivated meat across the classes they teach, plant biology professors to include a module on coaxing plant fats and proteins to behave like animal fats and proteins, and mechanical engineering professors to include readings and discussions of plant-based and cultivated meat production. With just a few motivated students, a campus can become a hub of alt protein innovation.

They can also convince professors to dedicate parts of their labs to alt meats and to apply for government grants. Some of today's leading alt protein researchers, including UCLA's Amy Rowat and the Technion's Shulamit Levenberg, entered the field because students asked them to.

Cultivated meat pioneer Natalie Rubio approached David Kaplan about supervising her PhD in biomedical engineering and cellular agriculture, which she completed in 2021, the third in the world.

In January 2025, Natalie told me about her happy incredulity at the field's progress since then. She shared that year after year, her optimism grows, thanks to cheaper media, AI, automation, and broader advances in tissue engineering. But the source of her greatest hope lies with students, who bring constant fresh energy, new tools, and ideas we haven't yet imagined.

My pride and joy at GFI is our university chapter program, the Alt Protein Project, which includes more than 75 university chapters worldwide. The program was designed by GFIer Amy Huang and is run by Amy and her colleague Nathan Ahlgrim. It's been successful beyond my wildest hopes. Everyone at GFI sets measurable goals, and we use Google's "objectives and key results" system to track our progress. Our student chapters are required to do the same, and they love it; their enthusiasm is infectious. One highlight of my 2025 was the three conferences I mentioned in the intro to this section; all three were organized by our student chapters. All three of these events were focused on deep science.

Ten years earlier, there had never been a plant-based or cultivated meat conference, and there was a sum of zero researchers on the planet receiving public funds to research alt meats. Less than 10 years later, there were three conferences at elite universities over just a few months, each conference featuring a back-to-back lineup of elite, publicly-funded alternative meat scientists. Honestly, my mind was blown.

AI adds another exciting layer: Google President of Global Affairs Kent Walker points to how Google's AlphaFold model has allowed scientists to understand more than 200 million proteins. Google Deepmind co-founder Demis Hassabis said that with protein folding, his team accomplished in one year what would have taken a billion years in a lab, "accurately predicting the 3D shape of proteins, critical for disease understanding and drug discovery."

These developments can be game-changing in our efforts to design plant-based and cultivated meat that mimics the texture, flavor, and functional properties of animal-based proteins. I could also imagine AI predicting modifications in proteins that improve their ability to hold water or fat. For cultivated meat, AI can assist in developing signaling molecules or enzymes that better facilitate muscle cell growth, optimize bioprocess efficiencies or improve scaffolding materials. The opportunities appear endless.

If you're a scientist or engineer (or open to becoming one), this is your moment. The challenges are real, the science is cutting-edge, and so is the potential for extraordinary impact.

Public Service to Accelerate Science and Scaling

GFI's number one goal is to create a maximally robust alt meats scientific ecosystem. Our global battle cry is that governments should be all-in on funding the transition. That's because governments fund the vast majority of the world's open access science, and governments are most responsible for helping global industries scale.

Government partnership with science and industry has been critical to many of the world's most transformative scientific endeavors. At GFI, we work with policymakers in every area of the world where we have an affiliate. In the US Congress, we've helped pass good bills and block bad ones. We secured alternative proteins funding in the USDA budget and helped to ensure that alt meats were included among White House strategies on advanced manufacturing and the American bioeconomy. Similar progress is underway all over the world.

Former National Science Foundation director Sethuraman Panchanathan was appointed by President Trump and confirmed 99-0 by the Senate. He often emphasizes that science and innovation have long enjoyed bipartisan support. Under his leadership, NSF funded the first cultivated meat research in the United States.

We need champions in legislative offices, and we need champions serving as program managers at the National Science Foundation and departments of defense, energy, and agriculture. Another great option for scientists: an AAAS or Horizon Institute fellowship in a legislative office or key agency.

Serving in government is one of the most powerful ways to accelerate progress on alternative proteins, whether you shape policy, direct funding, or build cross-party alliances. From Capitol Hill to agencies like USDA and NSF to international diplomacy, public service offers real influence. If you pursue this path (or you're there now), we'd love to work with you.

Academia: From Papers to Policy

Academic research provided the foundation for most of the information and analysis in this book. Academics write the papers that shape policy, present at conferences, and advise governments, nonprofits, and multilateral organizations. Their influence reaches every sector critical to the success of alternative proteins.

Here are two examples: research from a coalition of academics identifying alt proteins as a promising innovation to fight hunger and malnutrition (Innovation Commission for: Climate Change, Food Security, Agriculture) and the World AgriFood Innovation Conference in Beijing, hosted by China Agricultural University. The Innovation Commission work was led by University of Chicago economist Michael Kremer and Notre Dame agricultural economist Paul Winters. The WAFI event drew more than 1,000 attendees, including top officials from China's ministries for nutrition and agriculture. The entire event was the brainchild of professor Shenggen Fan, an agricultural economist who advises the Chinese government.

Academia produces research that drives public understanding, influences global agendas, and guides decision-makers. Despite the transformational promise of alt meats, the field remains vastly underrepresented in academia; there are thousands of academics working on poverty and malnutrition, climate and biodiversity, and global health. Many are championing solutions, but very few are connecting the dots from these global challenges to alternative proteins as a critical intervention.

Among climate-focused academics in particular, one specific need was pointed out to me by my friend Gabrielle Dreyfus, who serves as chief scientist at the Institute for Governance and Sustainable Development. She told me that the language of the climate world is Integrated Assessment Models. For alt meats to be thoroughly integrated into climate policy discussions, we're going to need the modeling community to engage much more deeply, building cost curves and running scenarios.

If you have this skill set, please know that this is a screaming need.

Whether you're an established scholar or just beginning your career, focusing some or all your work on the promise of alternative proteins could tip the scales toward a more sustainable, secure, and just food future.

Private Sector Pathways: From Startups to Incumbents

At the end of the day, alternative meats will scale through the private sector. As cultivated meat companies have struggled to raise money and as plant-based meat companies have struggled with sales, some have suggested that the startup craze was too early, that the tech was just not ready. I disagree. Recall the nearly 500 car companies that were formed between 1899 and 1908, when Henry Ford introduced the Model T. Most companies in any industry fail, but an industry can't succeed without its pioneers and a healthy dose of competitive encouragement toward innovation.

Plus, startups put alt meats on the radar of both governments and the major food and meat companies, and startups continue to drive excitement and innovation throughout the sector. As Tyson Foods' former sustainability VP Justin Whitmore said at a GFI event, "We don't want to be disrupted; we want to be the disruption."

Alt meats represent a multi-trillion-dollar opportunity for savvy entrepreneurs and investors. I'm convinced that there are ample white spaces in both plant-based and cultivated meat, some of which I've flagged throughout this book. I'd like nothing more than to see some of them taken up.

Business students often ask me how they can get a job in alternative proteins. My advice: If you're not a serial entrepreneur with a reasonable shot at creating a successful plant-based or cultivated meat research-focused startup, apply to work at a major food or meat company. We need more people working for Tyson, Cargill, JBS, Nestlé, and ADM who understand and believe in the alt meats vision. These companies will be essential to alt protein's long-term success. They have the infrastructure, capital, and marketing muscle to bring new products to scale. Their understanding of consumers and supply chains is unmatched. There is no better way to understand the food and meat industries than working at one of the biggest players.

Collaboration between startups and incumbents can also unlock even faster progress. Entrepreneurs benefit from access to resources, mentorship, and market knowledge, while established companies gain agility and fresh ideas. These partnerships can accelerate alt proteins from niche to mainstream. Beyond Meat founder and CEO Ethan Brown told me how helpful Tyson was to Beyond Meat's early growth, and John Randall Tyson told me that his father, Tyson board chair John Tyson, genuinely loved the product and aligned with the alt meats vision. That kind of enthusiasm matters.

Finally, entrepreneurs and business leaders can be powerful advocates. Their real-world experience makes them persuasive to policymakers across the political spectrum. They can make the case most authoritatively that alt meats aren't just good for the planet, they're also good business: creating jobs, revitalizing rural economies, and bolstering food security.

The Influence Network: Nonprofits and Think Tanks

Scientists do the bench work. Academics build the case for support. Governments fund and regulate. But it's often nonprofits, other cause-based organizations, and think tanks that connect the dots and drive momentum. These organizations convene experts, translate science into policy briefs, and push governments to act. They shape public and political discourse, influencing what gets prioritized by researchers, policymakers, journalists, and funders.

In chapter 1, I relied heavily on reports from the United Nations Food and Agriculture Organization (FAO). In chapter 2, I relied heavily on research conducted by the World Resources Institute and FAO. In chapters 3 and 4, I relied on reports from the World Health Organization and UNEP. In chapters 9 and 10, I relied on work by the Center for Strategic and International Studies. These six chapters would have been far less effective if not for the research and reports produced by these impressive and influential organizations.

Alt meats is still very early in its trajectory to the mainstream. We need far more of this research and these reports. Alt proteins should be as obvious a solution for climate as energy transition, which receives thousands of times more attention in the ecosystem of cause-based organizations. And support for alt proteins should be a key policy priority for many more of the vast ecosystem of organizations that work on environmental issues, hunger and malnutrition, and global health.

So here's my advice: If you work at any of these organizations, stay where you are, and work to incorporate alternative proteins into the work of your organization. Take it on as your personal priority. Use conferences, white papers, and media appearances to push it forward.

And if you're looking to contribute and have the appropriate skills, joining one of these cause-based organizations or an influential think tank may be one of the most impactful career decisions you can make.

Journalists: Storytellers That Change the World

Journalists are storytellers. They shape both how we understand the world and what we see as the range of tractable solutions to key global challenges. That's why academics, industry leaders, and cause-based organizations spend so much energy engaging with them.

While still at university, scientists can write for science blogs and similar online outlets, to prepare for career paths that include working for a scientific news outlet like *New Scientist, Scientific American, Popular Mechanics, Nature,* or one of the other journals in the *Nature* portfolio. Two of the three earliest endorsements of government support for alt meats science came from *New Scientist* and *Nature Biotechnology* (the other is Ezra Klein's opinion piece in the *New York Times*). These endorsements were incredibly helpful to GFI in our work with governments.

You don't have to be a scientist, of course, and most journalists aren't: Writing from the environmental or health desk of an influential news outlet can have immense impact, moving alternative proteins into a space where it's seen as an essential part of policy-focused conversations across the range of issues that alt proteins address. The same goes for editorial boards and op-ed writers. Most major news outlets have run both countless stories and dozens of editorials and opinion pieces in support of energy transition, and that's been critical to creating so much global support for action. We need some of those for protein transition.

Alt proteins can help to solve some of the world's most intransigent challenges, and working as a journalist can be an extremely impactful way of making sure that scientists and policymakers are aware of it.

Philanthropy: Where Smart Strategy Meets Scalable Impact

Philanthropy funds the organizations that create the political will to address the world's really big problems. It enabled much of the work I cited in chapters 1 through 4, as well as the CSIS report that formed the base of chapters 9 and 10. Although governments are responsible for the bulk of the funding going into hunger alleviation, energy transition, and global health, nonprofit organizations convince governments to set these priorities, and those nonprofit organizations are funded by philanthropy.

Climate philanthropy remains a tiny slice of global giving. According to the ClimateWorks Foundation, philanthropists donated $885 billion to nonprofit work in 2023, and less than 2% of that went to addressing climate change. Food systems receive even less, because most climate funders don't yet see food as a lever with high-impact, scalable interventions. Still, what's been achieved with limited food and climate funding is remarkable; the progress on renewable energy and electric vehicles could not have happened without the philanthropic sector. Philanthropy could do the same for alt meats.

For anyone with the right skills, working as a program officer at a foundation or as an advisor to a philanthropic family office can be a highly impactful career choice.

GFI: Recognized for Global Impact

Charity evaluators conduct deeply researched analyses of nonprofit organizations, with the goal of understanding which groups deliver the most impact in the evaluated sector.

The main climate charity evaluator is Giving Green, which was founded by Dan Stein, who worked previously as an economist at the World Bank and then as chief economist for IDinsight, a consultancy that conducts impact analyses of global development interventions in Africa and Asia; they help governments, foundations, and nonprofits to maximize positive outcomes.

Giving Green describes itself as "a team of climate scientists, economists and impact evaluation experts with decades of experience working at the intersection of evidence-based policy and the environment" and explains: "We spend thousands of hours reviewing the studies, crunching the numbers and interviewing the experts, so you don't have to."

Giving Green has analyzed GFI's theory of change and operations worldwide and reports that we are among the most impactful climate charities in the world, a position that we've maintained each year since 2022 (every year we've been evaluated).

Their analysts cite our "successful track record, breadth of expertise, and strategic approach" and call us "a powerhouse in alternative protein thought leadership and action."

Read their entire report at GivingGreen.earth.

Everyone Everywhere:
What You Know Can Change the World

I covered just a small sampling of vocations that someone might choose, if they want to maximize the impact of their vocational life on behalf of protein transition. But of course, there are many, many more.

At this moment, you're probably one of the only people in your workplace, school, faith community, or local government who understands the full potential of alt meats. These ideas are still unfamiliar to almost everyone, including experts. Even journalists covering climate and public health almost always miss them. So it's up to you to spark awareness. One email or conversation could make a huge positive difference.

Here are a few ideas with impact:

- Urge your elected officials to support government funding for alternative meat research and science-based policy.
- Talk to science and engineering professors at your local university about adding alt meats to their curriculums.

- Reach out to reporters when stories on climate, pandemics, or biodiversity miss the alt meats angle; put them in touch with GFI.
- Share this book with friends, colleagues, or community leaders. Organize a book club to discuss it.
- If you are connected to a person or people of influence—e.g., a legislator, science or climate influencer, climate or global health philanthropist—please talk with them about alt meats and consider giving them this book.
- If you give philanthropically, consider including GFI in your portfolio. Please see the box in this section containing a third-party validation of our impact.
- If you support environmental, global health, or food systems non-profits, encourage them to incorporate alternative proteins into their work.
- Sign up for GFI's e-news at gfi.org/newsletters to stay on top of the ecosystem.

Your efforts can tip the balance on climate change, biodiversity loss, hunger, malnutrition, AMR, pandemic risk, and animal protection. Whether you spark a scientist's curiosity, shift a policymaker's priorities, or convince a climate reporter to write a story, your actions can make a meaningful difference.

The Next Ten Years: Fixing the Hole in the Ship's Hull

There's a common view in climate circles that the next decade is *the* critical decade for climate and biodiversity. While that's true, if we continue to ignore meat's contribution to climate change and deforestation, everything we do in the next ten years will not be enough: Our climate will be unlivable, and there will be no forests left.

Recall from chapter 2: According to WRI's analysis (alongside the World Bank, the UN Environment Programme, and the UN Development Programme), on our current trajectory, agriculture will produce 15 billion metric tons of carbon—almost four times what's required to meet climate goals—and wipe out all of the world's forests and savannas by 2050.

So far, nothing we've tried has meaningfully slowed meat's contribution to climate change, and that despite clarion calls at the highest levels, going back decades (e.g., *Livestock's Long Shadow* came out in 2006).

Alt proteins can cut those emissions by 11.9 billion metric tons at 90 percent adoption, which is more than 1.3 billion for every 10% increase. So every 10% increase in alt proteins has about as much climate benefit as a global shift to EVs and more than eliminating air travel altogether. Alt proteins can also spare at least 650 million hectares of land, rather than requiring an additional 3 billion hectares.

So yes, let's do what we can over the next ten years, but if we ignore what appears to be the most promising intervention in agriculture—alternative proteins—we will not meet climate, biodiversity, or deforestation goals.

Plus, food will be more expensive, lifesaving drugs will stop working, and pandemic risk will climb.

The world is frantically bailing water from a sinking ship. But there's a gaping hole below the waterline. If we don't fix it, the ship is going down—no matter how fast we bail.

Some Closing Thoughts: Here's the Assignment

Thank you for joining me on this journey.

In some ways, my work to turn alternative meats from a science project into a mainstay of the global food supply chain feels worlds apart from my first post-university job in the early 1990s, running a Catholic Worker homeless shelter for families and soup kitchen in inner-city Washington, D.C.

For six years, I cooked food for America's most down-and-out, helped single moms to navigate the welfare and D.C. school systems, slept in a walk-in closet in the basement, and drove a few times per week through the most violent areas of what was then the "murder capital of the world," looking for people who needed food, clothing, blankets, or just a friend.

We weren't trying to solve poverty; we were trying to offer succor in the form of food, clothing, a warm bed, and a sympathetic ear.

But for me, it's a straight line from feeding the hungry to advocating for food systems change that will make it possible for hungry people all over the world to afford nutritious food.

Hunger and malnutrition are the issues that shook me into consciousness when I was 13, the reason I adopted a plant-based diet at 18, the reason I studied agricultural economics in college, and the reason I spent six years working with homeless families and two years teaching in inner-city schools through Teach For America, trying to find a way out and up for poor kids.

The alleviation of suffering is my life's mission.

As global meat consumption continues its steady upward trajectory, I see food prices climbing, exacerbating hunger and malnutrition. I see forests razed, the climate heating up, oceans dying, critical medicines failing, and pandemic risk rising. Those who suffer most from these effects of the world's insatiable desire for meat are the people least responsible. And I see billions upon billions of suffering farm animals, also innocent.

I find it deeply motivating to know that, after decades of watching industrial meat consumption's inexorable rise, we have a solution that could reverse that trajectory.

The assignment? Making it happen, as fast as possible.

Let's get to work.

Acknowledgments

T his book would not have been possible without support and insights from an extraordinary number of people, who helped me understand the material and shape it into what I hope is a coherent whole.

Many thanks to my agent Stacey Glick, for helping me to hone my idea for a book into a proper proposal and then for guiding the project to its happy home at BenBella Books.

To Leah Wilson, Sarah Modlin, Glenn Yeffeth, and the entire Ben-Bella team for your enthusiasm and encouragement and for being such a pleasure to work with at every stage of the process. Special thanks to Sarah Avinger and Morgan Carr for designing such a striking cover and to Kim Broderick for the lovely page design and layout.

To Sarah Morgan and Greg Brown for your invaluable assistance helping to organize the information in the most coherent way possible, cutting superfluous words and information, and making the entire text better. And to Suzanne Van Arsdale, for volunteering to do the painstaking work of checking and formatting the endnotes; thank you.

To Ilya Sheyman, GFI's former CEO, who convinced me to write this book, kept me on track when deadlines slipped, and provided multiple

Acknowledgments

rounds of incisive, actionable feedback that improved the manuscript in countless ways.

To GFI's entire science team, who I leaned on over and over and over as I conceptualized and then wrote chapters 6, 7, and 11. Special thanks to Claire Bomkamp, Erin Rees Clayton, Simone Costa, Lucas Eastham, Faraz Harsini, Amy Huang, Seren Kell, Niki Mansukhani Kogar, Adam Leman, Maanasa Ravikumar, Priera Panescu Scott, and Elliot Swartz. I'm in awe of your brilliance and grateful to you for putting it to use on behalf of our mission.

To the leaders of GFI's organizations around the world whose incredible work inspires me every day: Nir Goldstein, Mitre Gosker, Gus Guadagnini, Kimiko Hong-Mitsui, Yeonjoo La, Alex Mayers, Sneha Singh, and Nigel Sizer. Thanks also to GFI Consultancy CEO Doris Lee. The progress discussed in chapters 6, 7, 10, and 11 are a testament to your hard and deeply strategic work.

Thanks to the glue of the Regional Leadership Team, Rachel Lichte and Brian Berry, who keep us strategically integrated and on track for maximum impact. Nothing is more important.

To my GFI partners Ashley Pittman and Shayna Fertig, for both extensive book research and endless support across everything we're doing together. You are both so wonderfully upbeat and such fun to work with, in addition to being crazy-competent. I could not have completed this book without you.

To GFI's board, for making all of our work possible: Vandhana Bala, Nabiha Basathia, Kathy Freston, Joan Gass, Anand Gopal, and board chair Cameron Icard. I love working with all of you and cannot even begin to adequately express how much I value our partnership.

To all of GFI's teams around the world; you make every day a miracle.

So many thanks for reading the entire manuscript and offering insights and suggestions that made the book stronger throughout: Sophie Armour, Max Bazerman, Sarah David, Stewart David, Robert Gordon, Susan Haltman, Ryan Huling, Mikko Jarvenpaa, and Sheila Voss.

Acknowledgments

Thanks also to Stewart for shepherding the organization through its infancy and adolescence as our board chair for so many years; I don't know what GFI would have done without you.

For your invaluable assistance on key sections of the book, I'd like to thank Chris Bryant, Caroline Cotto, Andrew deCoriolis, Gabrielle Dreyfus, Max Elder, Grant Gordon, Michael Greger, Ryan Huling, Sarah LaHaye, Ann Linder, Raychel Santo, Liz Specht, Audrey and Matt Spence, Kent Walker, Zak Weston, and Amy Williams. The book is far more compelling due to your contributions.

Special thanks to Hannah Ritchie from Our World In Data (OWID), who is incredibly busy and even more talented, and who repeatedly pulled together and then helped me to understand critical data; I'm in awe of all you accomplish. And OWID is a goldmine of valuable information; what a service to the world.

I spent more than six months corresponding with more than 50 scientists who are working full-time on plant-based and cultivated meat, and I'm grateful to all of you, both for the engagement and, even more so, for your work to advance the science of plant-based and cultivated meat. Thanks in particular for your patience as I peppered you with more and more questions to Pat Brown, Allen Henderson, David Kaplan, and Mark Post.

To some critical partners and dear friends on the path to healthy and sustainable food systems: Parag Agarwal, Dan Altschuler, Rachel Atcheson, Sanah Baig, Josh Balk, Ryan Bethencourt, Clare Bland, Jack Bobo, Lewis Bollard, Vicky Bond, Anne Bordier, Lasse Bruun, Oliver Camp, Avery Cohn, Nick Cooney, Alan Darer, Varun Desphande, Rachel Dreskin, Ismahane Elouafi, Shenggen Fan, Sonalie Figueiras, Leah Garces, Morgan Gillespy, Chris Green, Craig Hanson, Andy Jarvis, Chris Jordan, Steve Jurvetson, David Kaplan, Jay Karandikar, Chris Kerr, Michael Kremer, Abhi Kumar, Sarah Lake, Chuck and Jennifer Laue, Kelly Levin, Sean McElwee, David Meyer, Ari Nessel, Dani Nierenberg, Kim Odhner, Richard Parr, Nigel Purvis, Jeremy Radachowsky, Milo Runkle,

Acknowledgments

Mike Ryan, Tim Searchinger, Nitin Sekar, Paul Shapiro, John and Timi Sobrato, Liz Specht, Jenny Stojkovic, Cleo Verkuijl, Diana Walsh, Rosie Wardle, Sir Robert Watson, David Welch, Suzy Welch, Paul Winters, and Costa Yiannoulis. You make food systems transformation work fun, and you inspire me beyond measure.

Special thanks to Caitlin Welsh, Joseph Majkut, and Zane Swanson from the Center for Strategic and International Studies. I am in your debt for seeing the importance of alternative proteins for the world and going first among think tanks in calling it out. I'm even more grateful that Caitlin agreed to write the foreword to this book; it's impossible to imagine anyone smarter or more thoughtful, and the book is much better due to your contribution. I am deeply inspired by your commitment to using your talents on behalf of life-saving policies, all over the world.

I am grateful every day to my mom and dad, Gus Friedrich and Rena Rae, for teaching me that the most important things in life are to stand up for those who are most vulnerable and to find a vocation that you enjoy. My mom passed away in May 2006, leaving a hole in my heart. She was enthusiastic about everything I did, so I know she would have enjoyed this book. I'm also grateful beyond words to Pastors Bud Dixen and David Klumpp, dear friends Kathy Boylan, Frank Cordaro, Bill and Sue Frankel-Streit, Liz McAlister, Art Laffin, Paul Magno, Patrick O'Neill and Mary Rider, Sisters Carol Gilbert and Ardeth Platte, and the many others who helped me to understand what it means to live a life of service. Although I fail constantly, I keep trying, as I have been for approaching 40 years.

Finally, for my life partner, co-conspirator in service, and moral compass, Alka Chandna. You support me when I need to be supported and challenge me when I need to be challenged; and you consistently help me to see the world in entirely new ways. We've been trying to make the world better together for a quarter century. I'm looking forward to the next quarter century or, AI willing, two.

Acknowledgments

And to our cats Rena, Tigger, and Angie—who bring us endless joy and remind us that every animal is unique, just like every human.

Of course, all errors of fact or analysis are mine alone, and I invite you to reach out to let me know about either. This book represents my current thinking on some topics that are—in my opinion—critical for global development, healthy ecosystems, global health, and animal protection. I'm not aware of another book that makes the case I've made in this book, and I'm hopeful that this will be the first among many, starting some conversations and advancing others. I'm sure I'll have my mind changed on many of the topics I'm discussing, and I look forward to that. I'll engage with critiques and document my latest thinking at MeatBook.org.

Notes

Introduction

1. "Per Capita Meat Consumption," IBISWorld, October 3, 2024, https://www.ibisworld.com/us/bed/per-capita-meat-consumption/4723/; "Meat and Dairy Production," Our World in Data, last modified December 2023, https://ourworldindata.org/meat-production ("The richer a country is, the more meat the average person typically eats . . . Overall, countries tend to shift upwards and to the right: getting richer and eating more meat"); "Total Meat Production," Our World in Data, accessed June 12, 2025, https://ourworldindata.org/grapher/meat-production-tonnes?tab=chart.

2. "The Most Popular American Dishes (Q3 2024)," YouGov, archived December 10, 2024, at https://web.archive.org/web/20241210074025/https://today.yougov.com/ratings/consumer/popularity/american-dishes/all.

Chapter 1

1. *Priority Innovations and Investment Recommendations for COP28* (Innovation Commission for Climate Change, Food Security and Agriculture, 2023), https://innovationcommission.uchicago.edu/wpcontent/uploads/2023/12/innovation_commission_-_cop28_innovation_cases_compiled.docx.pdf, 34.

2. Bruce Friedrich, "Market Forces and Food Technology Will Save the World," TEDx Sonoma County, November 2017, 17 min., 40 sec., https://www.ted.com/talks/bruce_friedrich_market_forces_and_food_technology_will_save_the_world_jan_2018.

3. "Share of Population Living in Extreme Poverty, World," Our World in Data, last updated October 9, 2023, https://ourworldindata.org/grapher/share-of-population-living-in-extreme-poverty-cost-of-basic-needs.

4. "Number of Moderately or Severely Food Insecure People," Our World in Data, last updated March 17, 2025, https://ourworldindata.org/grapher/number-of-people-moderately-or-severely-food-insecure.

Notes

5. Nicholas Kristof, "Save the Darfur Puppy," *New York Times*, May 10, 2007, https://www.nytimes.com/2007/05/10/opinion/10kristof.html.
6. FAO, IFAD, UNICEF, WFP, and WHO, *The State of Food Security and Nutrition in the World 2024*, Table 6.
7. FAO, IFAD, UNICEF, WFP, and WHO, *The State of Food Security and Nutrition in the World 2024*, Table 5.
8. Pierre J. Gerber, Henning Steinfeld, Benjamin Henderson, Anne Mottet, Carolyn Opio, Jeroen Dijkman et al., *Tackling Climate Change Through Livestock: A Global Assessment of Emissions and Mitigation Opportunities* (Food and Agriculture Organization of the United Nations, 2013), https://www.fao.org/4/i3437e/i3437e.pdf, 1.
9. Christina Pazzanese, "Amartya Sen's Nine-Decade Journey From Colonial India to Nobel Prize and Beyond," *The Harvard Gazette*, June 3, 2021, https://news.harvard.edu/gazette/story/2021/06/tracing-amartya-sens-path-from-childhood-during-the-raj-to-nobel-prize-and-beyond/.
10. Aditya Chakrabortty, "Secret Report: Biofuel Caused Food Crisis," *Guardian*, July 3, 2008, https://www.theguardian.com/environment/2008/jul/03/biofuels.renewableenergy.
11. George Monbiot, "Credit Crunch? The Real Crisis Is Global Hunger. And If You Care, Eat Less Meat," *Guardian*, April 14, 2008, https://www.theguardian.com/commentisfree/2008/apr/15/food.biofuels.
12. "Cereals Allocated to Food, Animal Feed and Fuel, World," Our World in Data, last updated March 17, 2025, https://ourworldindata.org/grapher/cereal-distribution-to-uses; "Share of Cereals Allocated to Animal Feed," Our World in Data, last updated March 17, 2025, https://ourworldindata.org/grapher/share-cereals-animal-feed.
13. "Wheat Exports From Ukraine and Russia in Perspective, 2021," Our World in Data, last updated March 17, 2025, https://ourworldindata.org/grapher/wheat-exports-ukraine-russia-perspective.
14. Monbiot, "Credit Crunch?"
15. *Global Innovation Needs Assessments: Protein Diversity*, (ClimateWorks Foundation and U.K. Foreign, Commonwealth & Development Office, 2021), https://www.climateworks.org/wp-content/uploads/2021/11/GINAs-Protein-Diversity.pdf; Marta Kozicka, Petr Havlík, Hugo Valin, Eva Wollenberg, Andre Deppermann, David Leclère et al., "Feeding Climate and Biodiversity Goals With Novel Plant-Based Meat and Milk Alternatives," *Nature Communications* 14, article no. 5316 (2023), https://doi.org/10.1038/s41467-023-40899-2.
16. "Alternative Proteins for Climate, Hunger, and Global Health," Baku, Azerbaijan, hosted by CGIAR and FAO at the 2024 United Nations Climate Change Conference (COP29), Food & Agriculture Pavilion, November 14, 2024, 1:30:15, https://youtu.be/q_30cep7FoM.
17. "Nature's Fynd Awarded $4.76 Million Grant To Advance its Breakthrough Fermentation Technology To Support Small Farming Households," Nature's Fynd, April 26, 2022, https://www.naturesfynd.com/press-release/bmgf-grant.
18. "Cellular Agriculture: A Global Food-Security Solution," Good Food Institute, March 15, 2018, https://gfi.org/blog/cellular-agriculture-a-global-food-security/.
19. "Cost-Effectiveness of 4 Specialized Nutritious Foods in the Prevention of Stunting and Wasting in Children Aged 6–23 Months in Burkina Faso: A Geographically Randomized Trial," *Current Developments in Nutrition* 4, no. 2 (2020): nzaa006, https://doi.org/10.1093/cdn/nzaa006. ("Since the [corn and soy blend] w/oil arm was not less effective than any of the

Notes

other foods for either stunting or wasting prevention, and was the least expensive, it was the most cost-effective product in this study . . . The results of this effectiveness trial are one more case in which the provision of animal-sourced foods were not more effective in preventing stunting and wasting than the provision of non–animal-sourced foods.")

20. "Climate Change: Overview," World Bank Group, last updated April 28, 2025, https://www.worldbank.org/en/topic/climatechange/overview.
21. Goodwin, "The Global Benefits of Reducing Food Loss and Waste, and How to Do It."

Chapter 2

1. *Protein Diversity*, (ClimateWorks Foundation and U.K. Foreign, Commonwealth & Development Office, 2021). (analysis based on 50% adoption)
2. Mark Rober, "Feeding Bill Gates a Fake Burger (To Save the World)," February 12, 2020, YouTube, 16 min., 38 sec., https://youtu.be/-k-V3ESHcfA.
3. Bill Gates, "Introducing the Green Premiums," GatesNotes, September 28, 2020, https://www.gatesnotes.com/Introducing-the-Green-Premiums.
4. António Guterres, "Secretary-General's statement on the IPCC Working Group 1 Report on the Physical Science Basis of the Sixth Assessment," United Nations, August 9, 2021, https://www.un.org/sg/en/content/sg/statement/2021-08-09/secretary-generals-statement-the-ipcc-working-group-1-report-the-physical-science-basis-of-the-sixth-assessment.
5. Secretary-General's opening remarks at press conference on climate António Guterres, "Secretary-General's Opening Remarks at Press Conference on Climate," United Nations, July 27, 2023, https://www.un.org/sg/en/content/sg/speeches/2023-07-27/secretary-generals-opening-remarks-press-conference-climate.
6. 'Humanity has opened the gates of hell,' UN Secretary-General says of climate urgency Stephanie Ebbs, "'Humanity Has Opened the Gates of Hell,' UN Secretary-General Says of Climate Urgency," *ABC News*, September 20, 2023, https://abcnews.go.com/International/humanity-opened-gates-hell-secretary-general-climate-urgency/story?id=103356944.
7. Creating a Sustainable Food Future (World Resources Institute, 2019).
8. Marco Springmann, Michael Clark, Daniel Mason-D'Croz, Keith Wiebe, Benjamin Leon Bodirsky, Luis Lassaletta et al., "Options for Keeping the Food System Within Environmental Limits," *Nature* 562:519-525. (2018), https://doi.org/10.1038/s41586-018-0594-0 (predicts an increase in agricultural emissions of 88% between 2010 and 2050; also finds that 72–78% of those emissions are attributable to farming animals).
9. "Total Meat Production," Our World in Data, last updated March 17, 2025, http://ourworldindata.org/grapher/meat-production-tonnes (that's the 2006 figure; now it's over 370 million metric tons).
10. William R. Sutton, Alexander Lotsch, and Ashesh Prasann, *Recipe for a Livable Planet: Achieving Net Zero Emissions in the Agrifood System*, Agriculture and Food Series (World Bank, 2024), http://hdl.handle.net/10986/41468; M. Crippa, E. Solazzo, D. Guizzardi, F. Monforti-Ferrario & A. Leip, "Food Systems Are Responsible for a Third of Global Anthropogenic GHG Emissions," *Nature Food* 2 (2021): 198-209, https://doi.org/10.1038/s43016-021-00225-9.
11. Whitney Bauck, "'Food Is Finally on the Table': Cop28 Addressed Agriculture in a Real Way," *Guardian*, December 17, 2023, https://www.theguardian.com/environment/2023/dec/17/cop28-sustainable-agriculture-food-greenhouse-gases.
12. "Global Fossil Fuel Consumption," Our World in Data, last updated June 20, 2024, https://

Notes

ourworldindata.org/grapher/global-fossil-fuel-consumption; *Phasing Down or Phasing Up? Top Fossil Fuel Producers Plan Even More Extraction Despite Climate Promises: Production Gap Report 2023* (Stockholm Environment Institute, Climate Analytics, E3G, International Institute for Sustainable Development, and United Nations Environment Programme, 2023) https://doi.org/10.51414/sei2023.050, 2, 35 ("Taken together, government plans and projections would lead to an increase in global coal production until 2030, and in global oil and gas production until at least 2050").

13. *Decarbonizing Hard-to-Abate Sectors with Renewables: Perspectives for the G7* (International Renewable Energy Agency, 2024), https://www.irena.org/-/media/Files/IRENA/Agency/Publication/2024/Apr/IRENA_G7_Decarbonising_hard_to_abate_sectors_2024.pdf; Michael Liebreich, "Liebreich: Net Zero Will Be Harder Than You Think – And Easier. Part I: Harder," *BloombergNEF*, September 6, 2023, https://about.bnef.com/blog/liebreich-net-zero-will-be-harder-than-you-think-and-easier-part-i-harder/; Steven Grattan, "World Far Off Track on Pledges to End Deforestation by 2030—Report," *Reuters*, October 24, 2023, https://www.reuters.com/business/environment/world-far-off-track-pledges-end-deforestation-by-2030-report-2023-10-23/ (The world is moving in the wrong direction on deforestation, too, but there are bright spots, including Brazil, which is making some admirable progress on the issue).

14. Hannah Ritchie, Pablo Rosado, and Max Roser, "Greenhouse Gas Emissions," Our World in Data, last updated January 2024, https://ourworldindata.org/greenhouse-gas-emissions (2023: 74.89, 19.54, 5.57); "Greenhouse Gas Emissions by Gas, World, 1850 to 2023," Our World in Data, last updated November 21, 2024, https://ourworldindata.org/grapher/methane-emissions (methane is up 23 % and N2O is up 22%, 2003 to 2023).

15. Greenhouse Gas Emissions, OWID, (2023: 74.89, 19.54, 5.57).

16. Eric Toensmeier, "Are Livestock Feed Additives the Future or Folly?," Project Drawdown, August 1, 2024, https://drawdown.org/insights/are-livestock-feed-additives-the-future-or-folly; Claudia Arndt, Alexander N. Hristov, William J. Price, and Zhongtang Yu, "Full Adoption of the Most Effective Strategies to Mitigate Methane Emissions by Ruminants Can Help Meet the 1.5 °C Target by 2030 but Not 2050," *Proceedings of the National Academy of Sciences* 119, no. 20: e2111294119, https://doi.org/10.1073/pnas.2111294119 (a team of 24 agricultural scientists reviewed more than 400 studies and found that methane mitigation from 100% adoption of the most effective strategies would still be significantly outweighed by the increase in ruminant numbers, see figure S4 in Appendix 1).

17. Food and Agriculture Organization of the United Nations, *Livestock in the Balance* (Rome, 2009), http://fao.org/4/i0680e/i0680e.pdf (The global population of cattle will increase from 1.5 billion to 2.6 billion and that of goats and sheep from 1.7 billion to 2.7 billion between 2000 and 2050).

18. Anne Mottet, Cees de Haan, Alessandra Falcucci, Giuseppe Tempio, Carolyn Opio, and Pierre Gerber, "Livestock: On Our Plates or Eating at Our Table? A New Analysis of the Feed/Food Debate," *Global Food Security* 14 (2017): 1, https://doi.org/10.1016/j.gfs.2017.01.001 (finds that between 7 and 13% of cattle are on feedlots; the remaining 87–93% are on range).

19. C. Alan Rotz, Senorpe Asem-Hiablie, Sara Place, and Greg Thoma, "Environmental Footprints of Beef Cattle Production in the United States," *Agricultural Systems* 169 (2019): 1, https://doi.org/10.1016/j.agsy.2018.11.005, table 2 (also confirmed independently by D. R. Brown, who writes "I ran some numbers for US Beef Feedlots based on USDA researcher Al Rotz's numbers: US beef feedlots make 10% of US beef cattle methane emissions per the

Notes

USDA"); Arndt et al., "Strategies to Mitigate Methane Emissions by Ruminants" (a team of 24 agricultural scientists reviewed more than 400 studies and found that methane mitigation from 100% adoption of the most effective strategies would still be significantly outweighed by the increase in ruminant numbers, see figure S4 in Appendix 1).

20. Eric Toensmeier, "Are Livestock Feed Additives the Future or Folly?," Project Drawdown, August 1, 2024, https://drawdown.org/insights/are-livestock-feed-additives-the-future-or-folly.

21. William R. Sutton, Alexander Lotsch, and Ashesh Prasann, *Recipe for a Livable Planet: Achieving Net Zero Emissions in the Agrifood System*, Agriculture and Food Series (World Bank, 2024) http://hdl.handle.net/10986/41468, Table 3.1, 72.

22. *Cultivating Change: A Collaborative Philanthropic Initiative to Accelerate and Scale Agroecology and Regenerative Approaches* (Global Alliance for the Future of Food, 2023 updated May 2024), https://futureoffood.org/wp-content/uploads/2024/05/GA_CultivatingChange_Report_052124.pdf; *Annex to Cultivating Change: Accelerating and Scaling Agroecology and Regenerative Approaches A Philanthropic Theory of Transformation* (Annex to Cultivating Change: Accelerating and Scaling Agroecology and Regenerative Approaches A Philanthropic Theory of Transformation, 2024), https://futureoffood.org/wp-content/uploads/2024/01/annex-cultivating-change2.pdf (the researchers report that current funding for regenerative agriculture totals about $44 billion per year, including $690 million from philanthropy (slide 21); that total number must scale to $4.3 trillion over 10 years, or $430 billion per year, which GAFF notes is less than 5% of the external costs of our food system—$12 trillion per year in health, environmental, and social costs annually, according to FAO analysis).

23. Tara Garnett, Cécile Godde, Adrian Muller, Elin Röös, Pete Smith, Imke de Boer et al., *Grazed and Confused? Ruminating on Cattle, Grazing Systems, Methane, Nitrous Oxide, the Soil Carbon Sequestration Question – and What It All Means for Greenhouse Gas Emissions* (Food Climate Research Network, Oxford Martin Programme on the Future of Food, and the Environmental Change Institute, University of Oxford, 2017), https://tabledebates.org/sites/default/files/2022-04/fcrn_gnc_report.pdf.

24. Jonathan Foley, "Greenwashing and Denial Won't Solve Beef's Enormous Climate Problems," Project Drawdown, October 30, 2024, https://drawdown.org/insights/greenwashing-and-denial-wont-solve-beefs-enormous-climate-problems.

25. Richard Waite, Jessica Zionts, and Clara Cho, *Toward "Better" Meat? Aligning Meat Sourcing Strategies With Corporate Climate and Sustainability Goals* (World Resources Institute, 2024), https://files.wri.org/d8/s3fs-public/2024-04/toward-better-meat_0.pdf, Box 3.

26. *Agriculture and Conservation: Living Nature in a Globalised World*, IUCN Flagship Report Series (International Union for Conservation of Nature and Natural Resources, 2024), https://doi.org/10.2305/AMHX3737.

27. Plant-based meat: Badding, Comparative Life Cycle Assessment (all plant-based numbers other than chicken land use); *Environmental Benefits of Alternative Proteins* (Good Food Institute, 2023), https://gfi.org/wp-content/uploads/2023/11/GFI23015_POL23010_Environmental-LCA-Factsheet_COP28-version.pdf (chicken land use). Plant-based chicken requires less land than plant-based pork because plant-based chicken does not use coconut oil, which is more land-intensive than other plant oils.

28. Cultivated meat: Pelle Sinke, Elliot Swartz, Hermes Sanctorum, Coen van der Giesen, and Ingrid Odegard, "Ex-Ante Life Cycle Assessment of Commercial-Scale Cultivated Meat Production in 2030," *The International Journal of Life Cycle Assessment* 28, no. 3 (2023): 234–254 (table 2); Pelle Sinke and Ingrid Odegard, *LCA of Cultivated Meat: Future*

Notes

Projections for Different Scenarios (CE Delft, February 2021), https://gfieurope.org/wp-content
/uploads/2022/04/CE_Delft_190107_LCA_of_cultivated_meat_Def.pdf (figure 14) (Cultivated chicken, pork, and beef comparisons are to most efficient farm animal systems, using
renewable energy for both systems).

29. *Protein Diversity* (ClimateWorks Foundation and U.K. Foreign, Commonwealth & Development Office, 2021), 15, FN4.
30. Benjamin Morach, Malte Clausen, Jürgen Rogg, Michael Brigl, Ulrik Schulze, Nico Deh-
nert et al., "The Untapped Climate Opportunity in Alternative Proteins," Food for Thought,
Boston Consulting Group, July 8, 2022, https://www.bcg.com/ja-jp/publications/2022
/combating-climate-crisis-with-alternative-protein (finding 2.2 Gt of mitigation at 22%
adoption; if emissions benefits grow in a linear fashion, this would compute to 5 Gt of mitigation at 50% adoption).
31. Marta Kozicka, Petr Havlík, Hugo Valin, Eva Wollenberg, Andre Deppermann, David Leclère et al., "Feeding Climate and Biodiversity Goals With Novel Plant-Based Meat and Milk
Alternatives," *Nature Communications* 14, article no. 5316 (2023), https://doi.org/10.1038
/s41467-023-40899-2.
32. Bill McKibben, "How Does Bill Gates Plan to Solve the Climate Crisis?," *New York Times*,
last updated March 9, 2021, https://www.nytimes.com/2021/02/15/books/review/bill-gates
-how-to-avoid-a-climate-disaster.html.
33. *Creating a Sustainable Food Future* (World Resources Institute, 2019).
34. Steinfeld et al., *Livestock's Long Shadow*, 7 (note that less than 1/200 of soy is fed to cattle for
beef; the vast majority is fed to chickens, farmed fish, and pigs); Hannah Ritchie, "Drivers
of Deforestation," Our World in Data, last updated May 2024, https://ourworldindata.org
/drivers-of-deforestation; Walter Fraanje and Tara Garnett, *Soy: Food, Feed, and Land Use
Change*, Foodsource: Building Blocks (Food Climate Research Network, University of Oxford, 2020), https://www.doi.org/10.56661/47e58c32.
35. Elizabeth Goldman, Sarah Carter, and Michelle Sims, "Fires Drove Record-breaking
Tropical Forest Loss in 2024," World Resources Institute, May 21, 2025, https://gfr.wri.org
/latest-analysis-deforestation-trends.
36. Steinfeld et al., *Livestock's Long Shadow*.
37. Ben Felder, "'I Didn't Think They Would Agree to Anything': Inside the Broken Negoti-
ations That Led to Oklahoma's 20-Year Case Against Several Poultry Companies," *Investigate Midwest*, January 8, 2024, https://investigatemidwest.org/2024/01/08/i-didnt-think
-they-would-agree-to-anything-inside-the-broken-negotiations-that-led-to-oklahomas
-20-year-case-against-several-poultry-companies/; Darryl Fears, "A Poultry Plant, Years
of Groundwater Contamination and, Finally, a Court Settlement," *Washington Post*, April
13, 2021, https://www.washingtonpost.com/climate-environment/2021/04/13/poultry-plant
-years-groundwater-contamination-finally-court-settlement/; Christopher Flavelle, "How
America's Diet Is Feeding the Groundwater Crisis," *New York Times*, December 24, 2023,
https://www.nytimes.com/interactive/2023/12/24/climate/groundwater-crisis-chicken
-cheese.html; Russ Bahorsky, "Study Reveals Air Pollution Inequities Linked to Industrial Swine Facilities Are Detectable from Space," January 28, 2025, University of Virginia,
https://as.virginia.edu/study-reveals-air-pollution-inequities-linked-industrial-swine-facilities
-are-detectable-space.
38. *Environmental Benefits of Alternative Proteins* (Good Food Institute, 2023); Sinke et al.,
"Life-Cycle Assessment in 2030."

Notes

39. Hannah Ritchie and Max Roser, "Fish and Overfishing," Our World in Data, last updated March 2024, https://ourworldindata.org/fish-and-overfishing.

40. "Fast Facts: U.S. Transportation Sector Greenhouse Gas Emissions 1990–2022," Environmental Protection Agency, May 2024, https://nepis.epa.gov/Exe/ZyPDF.cgi?Dockey=P101AKR0.pdf.

41. Ayurella Horn-Muller, "This Destructive Fishing Style Doesn't Just Harm Marine Life," *National Geographic*, January 17, 2024, https://www.nationalgeographic.com/environment /article/climate-change-bottom-trawling-fishing.

42. Jillian P. Fry, Nicholas A. Mailloux, David C. Love, Michael C. Milli, and Ling Cao, "Feed Conversion Efficiency in Aquaculture: Do We Measure It Correctly?," *Environmental Research Letters* 13: 024017, https://doi.org/10.1088/1748-9326/aaa273.

43. Lisa Jackson, "Soy Helped Build Aquaculture Into a Global Force. How Far Can It Take It?," Global Seafood Alliance, March 29, 2021, https://www.globalseafood.org/advocate /soy-helped-build-aquaculture-into-a-global-force-how-far-can-it-take-it/.

44. "Using Soy In Aquaculture Feeds," Fish Site, February 4, 2011, https://thefishsite.com /articles/using-soy-in-aquaculture-feeds; Morgan Cheatham and Tom D'Alfonso, "From Field to Fin: The Soy-Powered Future of Aquaculture," US Soybean Export Council, August 1, 2024, https://ussec.org/from-field-to-fin-the-soy-powered-future-of-aquaculture/.

45. *Stop Ghost Gear: The Most Deadly Form of Marine Plastic Debris* (World Wide Fund for Nature, 2020), https://www.worldwildlife.org/publications/stop-ghost-gear-the-most-deadly-form-of -marine-plastic-debris.

46. Emma Moore, Shannon Lyday, Jan Roletto, Kate Litle, Julia K Parrish, Hannah Nevins et al., "Entanglements of Marine Mammals and Seabirds in Central California and the North-West Coast of the United States 2001–2005," *Marine Pollution Bulletin* 58, no. 7 (2009): 1045–1051, https://doi.org/10.1016/j.marpolbul.2009.02.006.

47. Laurent Lebreton, Sarah-Jeanne Royer, Axel Peytavin, Wouter Jan Strietman, Ingeborg Smeding-Zuurendonk, and Matthias Egger, "Industrialised Fishing Nations Largely Contribute to Floating Plastic Pollution in the North Pacific Subtropical Gyre," *Scientific Reports* 12, article no. 12666 (2022), https://doi.org/10.1038/s41598-022-16529-0.

48. Global Innovation Needs Assessments: Food System Methane, Technical Report, April 2023, page 6, available: https://www.climateworks.org/ginas-methane/.

Chapter 3

1. Margaret Chan, "Antimicrobial Resistance in the European Union and the World," World Health Organization, March 14, 2012, https://www.who.int/director-general/speeches /detail/antimicrobial-resistance-in-the-european-union-and-the-world.

2. "World Leaders Commit to Decisive Action on Antimicrobial Resistance," World Health Organization, September 26, 2024, https://www.who.int/news/item/26-09-2024-world -leaders-commit-to-decisive-action-on-antimicrobial-resistance.

3. Ranya Mulchandani, Yu Wang, Marius Gilbert, and Thomas P. Van Boeckel, "Global Trends in Antimicrobial Use in Food-Producing Animals: 2020 to 2030," *PLOS Global Public Health* 3, no. 2 (2023): e0001305, https://doi.org/10.1371/journal.pgph.0001305 (article says that 99,502 metric tons are used in farm animals, which it says is 73% of global production; so total production would be 136,304 metric tons).

4. Christopher J. L. Murray, Kevin Shunji Ikuta, Fablina Sharara, Lucien Swetschinski,

Notes

Gisela Robles Aguilar, Authia Gray et al., "Global Burden of Bacterial Antimicrobial Resistance in 2019: A Systematic Analysis," *Lancet* 399, no. 10325 (2022): 626–655, https://doi.org/10.1016/S0140-6736(21)02724-0.

5. Mohsen Naghavi, Stein Emil Vollset, Kevin S. Ikuta, Lucien R. Swetschinski, Authia P. Gray, Eve E. Wool et al., "Global Burden of Bacterial Antimicrobial Resistance 1990–2021: A Systematic Analysis With Forecasts to 2050," *Lancet* 404, no. 10459 (2024): 1199–1226 https://doi.org/10.1016/S0140-6736(24)01867-1.

6. "2019 Antibiotic Resistance Threats Report," US Centers for Disease Control and Prevention, April 11, 2025, https://www.cdc.gov/antimicrobial-resistance/data-research/threats/index.html.

7. Sara Reardon, "Antibiotic Use in Farming Set to Soar Despite Drug-Resistance Fears: Analysis Finds Antimicrobial Drug Use in Agriculture Is Much Higher Than Reported," *Nature* (2023), https://doi.org/10.1038/d41586-023-00284-x.

8. In 2015, researchers from Princeton University, the International Livestock Research Institute, and the One Health Trust in Washington, D.C., estimated that roughly 63,151 metric tons of antibiotics were fed to farm animals in 2010, and they predicted that by 2030, that number would be up 67% to roughly 105,000 metric tons. Thomas P. Van Boeckel, Charles Brower, Marius Gilbert, Bryan T. Grenfell, Simon A. Levin, and Timothy P Robinson, "Global Trends in Antimicrobial Use in Food Animals," *Proceedings of the National Academy of Sciences* 112, no. 18 (2015): 5649–5654. By 2020, use had already reached 99,502 metric tons, so researchers increased their 2030 prediction to 107,472 tons. Mulchandani et al., "Global Trends in Antimicrobial Use in Food-Producing Animals: 2020 to 2030."

9. Samantha Serrano, "Aquaculture & Antimicrobial Resistance," One Health Trust, September 13, 2022, https://onehealthtrust.org/publications/infographics/aquaculture-antimicrobial-resistance/.

10. *Bracing for Superbugs: Strengthening Environmental Action in the One Health Response to Antimicrobial Resistance* (United Nations Environment Programme, 2023), https://www.unep.org/resources/superbugs/environmental-action, 38–41; Joy E.M. Watts, Harold J. Schreier, Lauma Lanska, and Michelle S. Hale, "The Rising Tide of Antimicrobial Resistance in Aquaculture: Sources, Sinks and Solutions," *Marine Drugs* 15, no. 6 (2017): 158, https://doi.org/10.3390/md15060158; Daniel Schar, Eili Y. Klein, Ramanan Laxminarayan, Marius Gilbert, and Thomas P. Van Boeckel, "Global Trends in Antimicrobial Use in Aquaculture," *Scientific Reports* 10 (2020): 21878, https://doi.org/10.1038/s41598-020-78849-3 ("aquaculture settings utilizing antimicrobials may serve as reservoirs for antimicrobial resistance genes, providing routes for human and animal exposure to antimicrobial resistant bacteria").

11. *Imported Seafood Safety: FDA Should Improve Monitoring of Its Warning Letter Process and Better Assess Its Effectiveness* (Government Accountability Office, GAO-21-231, March 19, 2021), https://www.gao.gov/assets/d21231.pdf; *Food Safety: FDA Should Strengthen Inspection Efforts to Protect the U.S. Food Supply* (Government Accountability Office, GAO-25-107571, January 8, 2025), https://www.gao.gov/assets/gao-25-107571.pdf (from 2018-2023, FDA's best year for overseas food facility inspections was shy of 10% of what the Food Safety Modernization Act requires).

12. Joanne C. Chee-Sanford, Roderick I. Mackie, Satoshi Koike, Ivan G. Krapac, Yu-Feng Lin, Anthony C. Yannarell et al., "Fate and Transport of Antibiotic Residues and Antibiotic Resistance Genes Following Land Application of Manure Waste," *Journal of Environmental*

Notes

Quality 38, no. 3 (2009): 1086–1108, https://doi.org/10.2134/jeq2008.0128; Schar et al., "Global Trends in Antimicrobial Use in Aquaculture"; Huibo Xin, Min Gao, Xuming Wang, Tianlei Qiu, Yajie Guo, and Liqiu Zhang, "Animal Farms Are Hot Spots for Airborne Antimicrobial Resistance," *Science of the Total Environment* 851, no. 1 (2022): 158050, https://doi .org/10.1016/j.scitotenv.2022.158050.

13. Miriam Reverter, Samira Sarter, Domenico Caruso, Jean-Christophe Avarre, Marine Combe, Elodie Pepey et al., "Aquaculture at the Crossroads of Global Warming and Antimicrobial Resistance," *Nature Communications* 11, article no. 1870 (2020), https://doi .org/10.1038/s41467-020-15735-6 ("About 80% of antimicrobials administered through feed to aquatic farmed animals disseminate to nearby environments (water and sediment) where they remain active for months at concentrations allowing selective pressure on bacterial communities and favouring AMR development").

14. Susanne A. Kraemer, Arthi Ramachandran, and Gabriel G. Perron, "Antibiotic Pollution in the Environment: From Microbial Ecology to Public Policy," *Microorganisms* 7, no. 6 (2019): 180, https://doi.org/10.3390/microorganisms7060180; Martin J. Blaser, "Antibiotic use and its consequences for the normal microbiome," *Science*, 352, no. 6285 (2016): 544–545, https:// doi.org/10.1126/science.aad9358.

15. Claas Kirchhelle, "Swann Song: Antibiotic Regulation in British Livestock Production (1953–2006)," *Bulletin of the History of Medicine* 92, no. 2 (2018): 317–350, https://doi .org/10.1353/bhm.2018.0029.

16. Emerging and Other Communicable Diseases: Antimicrobial Resistance, approved resolution of the Fifty-First World Health Assembly, 16 May 1998, https://iris.who.int/bitstream /handle/10665/79863/ear17.pdf.

17. *WHO Global Principles for the Containment of Antimicrobial Resistance in Animals Intended for Food: Report of a WHO Consultation With the Participation of the Food and Agriculture Organization of the United Nations and the [World Organization for Animal Health], Geneva, Switzerland 5–9 June 2000* (World Health Organization, 2000), https://iris.who.int/handle/10665/68931.

18. *WHO Global Strategy for Containment of Antimicrobial Resistance* (World Health Organization, January 1, 2001), https://www.who.int/publications/i/item/who-global-strategy-for -containment-of-antimicrobial-resistance.

19. Mary J. Gilchrist, Christina Greko, David B. Wallinga, George W. Beran, David G. Riley, and Peter S. Thorne, "The Potential Role of Concentrated Animal Feeding Operations in Infectious Disease Epidemics and Antibiotic Resistance," *Environmental Health Perspectives* 115, no. 2 (2007): 313–316, https://doi.org/10.1289/ehp.8837.

20. Jean Carlet, Peter Collignon, Don Goldmann, Herman Goossens, Inge C. Gyssens, Stephan Harbarth et al., "Society's Failure to Protect a Precious Resource: Antibiotics," *Lancet* 378, no. 9788 (2011): 369–371, https://doi.org/10.1016/s0140-6736(11)60401-7.

21. "At UN, Global Leaders Commit to Act on Antimicrobial Resistance," General Assembly of the United Nations, September 21, 2016, https://www.un.org/pga/71/2016/09/21 /press-release-hl-meeting-on-antimicrobial-resistance/.

22. "Stop Using Antibiotics in Healthy Animals to Prevent the Spread of Antibiotic Resistance," World Health Organization, November 7, 2017, https://www.who.int/en/news-room /detail/07-11-2017-stop-using-antibiotics-in-healthy-animals-to-prevent-the-spread-of -antibiotic-resistance.

23. "Ensuring Progress on Sustainable Access to Effective Antibiotics at the 2024 UN General

Notes

Assembly: A Target-Based Approach," *Lancet* 403, no. 10443 (2024): 2551–2564 (the scientists represent the United States, UK, South Africa, Ethiopia, Zambia, Tanzania, Canada, and India).

24. "Amid the Escalating Impact of Antimicrobial Resistance, the Global Leaders Group Calls on UN Member States to Take Bold and Specific Action," Global Leaders Group on Antimicrobial Resistance, April 2024, https://www.amrleaders.org/news-and-events/news/item/03-04-2024-amid-the-escalating-impact-of-antimicrobial-resistance-the-global-leaders-group-calls-on-un-member-states-to-take-bold-and-specific-action.

25. "Global Trends in Antimicrobial Use in Food Animals from 2017 to 2030," *Antibiotics*. 2020 Dec 17;9(12):918. doi: 10.3390/antibiotics9120918.

26. "Global Increase and Geographic Convergence in Antibiotic Consumption Between 2000 and 2015," *Proceedings of the National Academy of Sciences* 115, no. 15 (2017): E3463–E3470, https://doi.org/10.1073/pnas.1717295115; Eili Y. Klein, Isabella Impalli, Suprena Poleon, and Arindam Nandi, "Global Trends in Antibiotic Consumption During 2016–2023 and Future Projections Through 2030," *Proceedings of the National Academy of Sciences* 121, no. 49 (2024): e2411919121, https://doi.org/10.1073/pnas.2411919121.

27. Ramanan Laxminarayan, Precious Matsoso, Suraj Pant, Charles Brower, John-Arne Røttingen, Keith Klugman, and Sally Davies, "Access to Effective Antimicrobials: A Worldwide Challenge," *Lancet* 387, no. 10014 (2016): 168–175, https://doi.org/10.1016/S0140-6736(15)00474-2; Annie J. Browne, Michael G. Chipeta, Georgina Haines-Woodhouse, Emmanuelle P.A. Kumaran, Bahar H. Kashef Hamadani, Sabra Zaraa et al., "Global Antibiotic Consumption and Usage in Humans, 2000–18: A Spatial Modelling Study," *Lancet Planetary Health* 5, no. 12 (2021), https://doi.org/10.1016/S2542-5196(21)00280-1.

28. David Hyun, "World Health Organization Warns That Antibiotic Innovation Still Insufficient," Pew, July 15, 2024, https://www.pewtrusts.org/en/research-and-analysis/articles/2024/07/15/world-health-organization-warns-that-antibiotic-innovation-still-insufficient; Allan Coukell and Helen Boucher, "The Antibiotic Market Is Broken—and Won't Fix Itself," Pew, April 10, 2019, https://www.pewtrusts.org/en/about/news-room/opinion/2019/04/10/the-antibiotic-market-is-broken-and-wont-fix-itself.

29. *Executive Summary: Incentivising the Development of New Antibacterial Treatments: 2023 Progress Report By The Global AMR R&D Hub and WHO* (Global AMR R&D Hub and World Health Organization, 2023), https://cdn.who.int/media/docs/default-source/2021-dha-docs/incentivising-development-of-new-antibacterial-treatments-2023---exec-summary.pdf.

30. "Lack of Innovation Set to Undermine Antibiotic Performance and Health Gains," World Health Organization, June 2022, https://www.who.int/news/item/22-06-2022-22-06-2022-lack-of-innovation-set-to-undermine-antibiotic-performance-and-health-gains.

31. *Report of the Sixth Meeting of the Global Leaders Group on AMR* (One Health Global Leaders Group on Antimicrobial Resistance, March 2023), https://www.amrleaders.org/resources/m/item/report-of-the-sixth-meeting-of-the-global-leaders-group-on-amr.

32. AbdulRahman A. Saied, Deepak Chandran, Hitesh Chopra, Abhijit Dey, Talha B. Emran, and Kuldeep Dhama, "Cultivated Meat Could Aid in Reducing Global Antimicrobial Resistance Burden – Producing Meat Without Antibiotics as a Safer Food System for the Future," *International Journal of Surgery* 109, no. 2 (2023): 189–190, https://doi.org/10.1097/js9.0000000000000199; Cultivated Meat as a Tool for Fighting Antimicrobial Resistance, *Nature Food*, Oct 2022.

Notes

Chapter 4

1. *What's Cooking? An Assessment of Potential Impacts of Selected Novel Alternatives to Conventional Animal Products* (United Nations Environment Programme, December 2023), https:// www.unep.org/resources/whats-cooking-assessment-potential-impacts-selected-novel -alternatives-conventional.

2. Charles W Schmidt, "Swine CAFOs & Novel H1N1 Flu: Separating Facts from Fears," *Environmental Health Perspectives* 117, no. 9 (2009): A394–A401, https://doi.org/10.1289 /ehp.117-a394; "1880–1959 Highlights in the History of Avian Influenza (Bird Flu) Timeline," Centers for Disease Control and Prevention, April 30, 2024, https://www.cdc.gov /bird-flu/avian-timeline/1880-1959.html.

3. Dennis E. Showalter and John Graham Royde-Smith, "World War I: Killed, Wounded, and Missing," Britannica, last updated June 14, 2025, https://www.britannica.com/event/World -War-I/Killed-wounded-and-missing; Thomas A. Hughes and John Graham Royde-Smith, "World War I: Costs of the War," Britannica, last updated June 13, 2025, https://www .britannica.com/event/World-War-II/Costs-of-the-war.

4. Jeffery K. Taubenberger and David M. Morens, "The Pathology of Influenza Virus Infections," *Annual Review of Pathology: Mechanisms of Disease* 3, no. 1 (2008): 499–522, https://doi .org/10.1146/annurev.pathmechdis.3.121806.154316.

5. Emily Athens and Linda Qui, "Farm Animals Are Hauled All Over the Country. So Are Their Pathogens," *New York Times*, May 20, 2024, https://www.nytimes.com/2024/05/20 /health/livestock-disease-transport.html.

6. Schmidt, "Swine CAFOs & novel H1N1 flu," https://doi.org/10.1289/ehp.117-a394.

7. Maryn Mckenna, "Bird Flu Cost the US $3.3 Billion and Worse Could Be Coming," *National Geographic*, July 15, 2015, https://www.nationalgeographic.com/science/article/bird -flu-2 (cost to business); Sean E. Ramos, Matthew MacLachlan, and Alex Melton, *Impacts of the 2014-2015 Highly Pathogenic Avian Influenza Outbreak on the U.S. Poultry Sector*, LDPM-282-02 (US Department of Agriculture, Economic Research Service, December 20, 2017), https://www.ers.usda.gov/publications/pub-details?pubid=86281 (cost to the government).

8. *Avian Influenza: USDA Has Taken Actions to Reduce Risks but Needs a Plan to Evaluate Its Efforts* (Government Accountability Office, GAO-17-360, May 11, 2017), https://www.gao .gov/assets/gao-17-360.pdf.

9. Ignacio Mena, Martha I. Nelson, Francisco Quezada-Monroy, Jayeeta Dutta, Refugio Cortes-Fernández, J. Horacio Lara-Puente et al., "Origins of the 2009 H1N1 Influenza Pandemic in Swine in Mexico," *eLife* 5 (2016): e16777, https://doi.org/10.7554/eLife.16777.

10. Fatimah S Dawood, A. Danielle Iuliano, Carrie Reed, Martin I. Meltzer, David K. Shay, Po-Yung Cheng et al., "Estimated Global Mortality Associated With the First 12 Months of 2009 Pandemic Influenza a H1N1 Virus Circulation: A Modelling Study," *Lancet Infectious Diseases* 12, no. 9 (2012): 687-695, https://doi.org/10.1016/s1473-3099(12)70121-4.

11. Donald G. McNeil Jr., "U.S. Reaction to Swine Flu: Apt and Lucky," *New York Times*, January 1, 2010, https://www.nytimes.com/2010/01/02/health/02flu.html.

12. Dinah Henritzi, Philipp Peter Petric, Nicola Sarah Lewis, Annika Graaf, Alberto Pessia, Elke Starick et al., "Surveillance of European Domestic Pig Populations Identifies an Emerging Reservoir of Potentially Zoonotic Swine Influenza A Viruses," *Cell Host & Microbe* 28, no. 4 (2020): 614–627, https://doi.org/10.1016/j.chom.2020.07.006; "The Reasons Swine Flu Could Return," BBC, February 1, 2021, https://www.bbc.com/future /article/20210202-Swine-flu-why-influenza-in-pigs-could-cause-another-pandemic.

Notes

13. Brad Gilbertson and Kanta Subbarao, "Mammalian Infections with Highly Pathogenic Avian Influenza Viruses Renew Concerns of Pandemic Potential," *Journal of Experimental Medicine* 220, no. 8 (2023): e20230447, https://doi.org/10.1084/jem.20230447.
14. Gilbertson and Subbarao, "Mammalian Infections with Highly Pathogenic Avian Influenza."
15. Katie Gillespie, "Are We Subsidizing the Next Pandemic? How Government Payments to Big Poultry Threaten Public Health," Farm Forward, March 2025, https://www.farmforward.com/wp-content/uploads/2025/03/Bird-Flu-Brief-1.2.pdf.
16. Apoorva Mandavilli, "A Bird-Flu Pandemic in People? Here's What It Might Look Like," *New York Times*, last updated June 20, 2024, https://www.nytimes.com/2024/06/17/health/bird-flu-pandemic-humans.html.
17. Emily Anthes, "Bird Flu Is Still Causing Havoc. Here's the Latest," *New York Times*, December 15, 2023, https://www.nytimes.com/2023/12/15/science/birds-flu-h5n1.html; *Technical Report: June 2024 Highly Pathogenic Avian Influenza A(H5N1) Viruses* (US Centers for Disease Control and Prevention, June 5, 2024), https://www.cdc.gov/bird-flu/php/technical-report/h5n1-06052024.html.
18. *Technical Report: June 2024 Highly Pathogenic Avian Influenza A(H5N1) Viruses*, CDC.
19. Mandavilli, "A Bird-Flu Pandemic in People?"
20. Mandavilli, "A Bird-Flu Pandemic in People?"
21. David A. Kessler, "I Ran Operation Warp Speed. I'm Concerned About Bird Flu," *New York Times*, November 26, 2024, https://www.nytimes.com/2024/11/26/opinion/vaccine-bird-flu-pandemic.html; Zeynep Tufekci, "A Bird Flu Pandemic Would Be One of the Most Foreseeable Catastrophes in History," *New York Times*, November 29, 2024, https://www.nytimes.com/2024/11/29/opinion/bird-flu-pandemic.html.
22. *Report of the WHO/FAO/OIE Joint Consultation on Emerging Zoonotic Diseases* (Food and Agriculture Organization of the United Nations, World Health Organization, and World Organisation for Animal Health, 2004), https://iris.who.int/bitstream/handle/10665/68899/WHO_CDS_CPE_ZFK_2004.9.pdf.
23. Michael Greger, "Primary Pandemic Prevention," *American Journal of Lifestyle Medicine* 15, no. 5 (2021): 498–505, https://doi.org/10.1177/15598276211008134.

Chapter 5

1. The Simpsons, Lisa the Vegetarian (aired Oct. 15, 1995), quote available at https://www.imdb.com/title/tt0701158/quotes/?item=qt1929673.
2. Jay L. Zagorsky and Patricia K. Smith, "The Association Between Socioeconomic Status and Adult Fast-Food Consumption in the U.S.," *Economics & Human Biology* 27, Part A (2017): 12–25, https://doi.org/10.1016/j.ehb.2017.04.004; Ohio State University, "For Richer or Poorer, We All Eat Fast Food," ScienceDaily, May 4, 2017, www.sciencedaily.com/releases/2017/05/170504083222.htm.
3. Nicholas Bakalar, "Fast Food: It's What's for Dinner. And Lunch. And Breakfast," *New York Times*, October 22, 2018 https://www.nytimes.com/2018/10/22/health/fast-food-consumption.html (37% of all adults and 45% of younger adults).
4. We only have meat consumption statistics going back to 1961, when the world was eating 71 million metric tons of meat per year—across roughly 3 billion people. Hannah Ritchie, Pablo Rosado, and Max Roser, "Meat and Dairy Production," Our World in Data, last updated December 2023, https://ourworldindata.org/meat-production. Assuming the same per capita consumption, this would indicate that 42.6 MMT of meat were consumed in 1918.

Notes

This is conservative, since we know that meat consumption rises with income, and poverty rates globally fell from 59.6% to 50.24% during this time. Max Roser, "The Short History of Global Living Conditions and Why It Matters That We Know It," Our World in Data, last updated February 2024, https://ourworldindata.org/a-history-of-global-living-conditions. In the three decades after 1961, meat consumption rose by 31% to 33% per decade: 70.57 MMT in 1961; 103.59 in 1971; 137.77 in 1981; 181.05 in 1991.

5. Sarah A. Rydell, Lisa J. Harnack, J. Michael Oakes, Mary Story, Robert W. Jeffery, Simone A. French, "Why Eat at Fast-Food Restaurants: Reported Reasons Among Frequent Consumers," *Journal of the American Dietetic Association* 108, no. 12 (2008): 2066–70, https://doi.org/10.1016/j.jada.2008.09.008.

6. Jungwon Min, Lisa Jahns, Hong Xue, Jayanthi Kandiah, and Youfa Wang, "Americans' Perceptions about Fast Food and How They Associate with Its Consumption and Obesity Risk," *Advances in Nutrition* 9, no. 5 (2018): 590–601, https://doi.org/10.1093/advances/nmy032.

7. "Two-thirds of U.S. consumers say they are eating less meat," Sept 12, 2018, https://hub.jhu.edu/2018/09/12/consumers-cut-back-on-meat-consumption/.

8. Yasmin Tayag, "America Is Done Pretending About Meat," *Atlantic*, March 24, 2025, https://www.theatlantic.com/health/archive/2025/03/meat-boom-trump-rfk-jr/682150/.

9. Kim Severson, "Meat Is Back, on Plates and in Politics," *New York Times*, April 18, 2025, https://www.nytimes.com/2025/04/18/dining/meat-beef-restaurants-politics.html.

10. "Per Capita Meat Consumption," IBIS World, October 3, 2024, https://www.ibisworld.com/us/bed/per-capita-meat-consumption/4723/.

11. "USDA Agricultural Projections to 2033," US Department of Agriculture, February 2024, https://www.usda.gov/sites/default/files/documents/USDA-Agricultural-Projections-to-2033.pdf, 49.

12. National Center for Chronic Disease Prevention and Health Promotion (US) Office on Smoking and HealthCenters for Disease Control and Prevention (US), "Fifty Years of Change 1964–2014," in *The Health Consequences of Smoking—50 Years of Progress: A Report of the Surgeon General* (US Centers for Disease Control and Prevention, 2014), https://www.ncbi.nlm.nih.gov/books/NBK294310/.

13. *WHO Global Report on Trends in Prevalence of Tobacco Use 2000–2030* (World Health Organization, 2024), https://iris.who.int/bitstream/handle/10665/375711/9789240088283-eng.pdf.

14. "Sales at Record High, Americans View Meat as Part of a Healthy, Balanced Lifestyle Power of Meat Analysis Marks 20th Anniversary at Annual Meat Conference," Meat Institute, March 24, 2025, https://meatinstitute.org/press/sales-record-high-americans-view-meat-part-healthy-balanced-lifestyle-power-meat-analysis.

15. Mark Bittman, "What's Wrong With What We Eat?" TED, December 2007, 19 min., 46 sec., https://www.ted.com/talks/mark_bittman_what_s_wrong_with_what_we_eat ("This [image of cow] may be this year's version of this [image of nuclear explosion] . . . Because only once before has the fate of individual people and the fate of all of humanity been so intertwined. There was the bomb, and there's now. And where we go from here is going to determine not only the quality and the length of our individual lives, but whether if we could see the earth a century from now, we'd recognize it. It's a holocaust of a different kind . . .").

16. "Decadent-Sounding Descriptions Could Lead to Higher Consumption of Vegetables, Stanford Research Finds," Stanford Report, June 12, 2017, https://news.stanford.edu/2017/06/12/decadent-sounding-labeling-may-lead-people-eat-vegetables/.

Notes

17. "Obesity and Overweight," World Health Organization, May 7, 2025, https://www.who.int/news-room/fact-sheets/detail/obesity-and-overweight.
18. Cheryl D. Fryar, Margaret D. Carroll, and Joseph Afful, "Prevalence of Overweight, Obesity, and Severe Obesity Among Adults Aged 20 and Over: United States, 1960–1962 Through 2017–2018," US Centers for Disease Control and Prevention, last updated January 29, 2021, https://www.cdc.gov/nchs/data/hestat/obesity-adult-17-18/obesity-adult.htm.
19. John Stossel, 20/20, "Tug of War Over the Name 'Milk,'" aired March 24, 2000, on ABC.
20. "Definition & Facts for Lactose Intolerance," National Institutes of Health, https://www.niddk.nih.gov/health-information/digestive-diseases/lactose-intolerance/definition-facts.

Chapter 6

1. "Google Burger: Sergey Brin Explains Why He Funded World's First Lab-Grown Beef Hamburger," Guardian, August 5, 2023, 6 min., 4 sec., https://www.theguardian.com/science/video/2013/aug/05/google-burger-sergey-brin-lab-grown-hamburger.
2. Meat the Future, directed by Liz Marshall (LizMars Productions, 2020), https://meatthefuture.com/.
3. Mackenzie Battle, Michael Carter, Ankur Chaudhary, Alex Holst, Kimiko Hong-Mitsui, Yeonjo La et al., 2024 State of Global Policy: Public Investment in Alternative Proteins to Feed a Growing World (Good Food Institute, 2025), https://gfi.org/wp-content/uploads/2025/04/2024-State-of-global-policy-Public-investment-in-alternative-proteins-to-feed-a-growing-world.pdf; David Kaplan and Bruce Friedrich, "Fireside Chat Keynote - Tufts University Cellular Agriculture Innovation Day 2024," January 11, 2024, posted by February 9, 2024, YouTube, 32 min., 26 sec., https://youtu.be/iGzXAuiYKGU.
4. David Kaplan and Bruce Friedrich, "Fireside Chat Keynote - Tufts University Cellular Agriculture Innovation Day 2024," January 11, 2024, posted by February 9, 2024, YouTube, 32 min., 26 sec., https://youtu.be/iGzXAuiYKGU.
5. "Ali Khademhosseini, CEO," Terasaki Institute for Biomedical Innovation, accessed June 15, 2025, https://terasaki.org/institute/ali.
6. Mackenzie Battle, Claire Bomkamp, Michael Carter, Jessica Colley Clarke, Lucas Eastham, Liz Fathman et al., 2024 State of the Industry: Cultivated Meat, Seafood, and Ingredients (Good Food Institute, 2025), https://gfi.org/wp-content/uploads/2025/04/2024-State-of-the-Industry-Cultivated-meat-seafood-and-ingredients-GFI.pdf; Asia additions from GFI APAC scientist Maanasa Ravikumar.
7. Laura Pasitka, Merav Cohen, Avner Ehrlich, Boaz Gildor, Eli Reuveni, Muneef Ayyash et al., "Spontaneous Immortalization of Chicken Fibroblasts Generates Stable, High-Yield Cell Lines for Serum-Free Production of Cultured Meat," Nature Food 4, no. 1 (2022): 35–50. https://doi.org/10.1038/s43016-022-00658-w; "Hebrew University Study, Led By Believer Meats Founder, Unveils a Highly Efficient Process for the GMO-Free Production of Cultured Meat, Setting New Standards for Transparency in the Field," press release, Protein Report, January 5, 2023, https://www.proteinreport.org/newswire/hebrew-university-study-led-believer-meats-founder-unveils-highly-efficient-process-gmo/; Elaine Watson, "The Most Comprehensive Study on Cultivated Meat Production Published to Date," Believer Meats Data Shines Light on Cell Cultured Meat Metrics," Food Navigator, last updated May 13, 2024, https://www.foodnavigator-usa.com/Article/2023/01/11/Believer-Meats-study-in-Nature-Food-shines-light-on-cultivated-meat-metrics/.
8. "Quality US Patents Drive Our Economy and Solve World Problems," Director's Blog,

Notes

US Patent and Trademark Office, December 9, 2024, https://www.uspto.gov/subscription-center/2024/quality-us-patents-drive-our-economy-and-solve-world-problems; David Hsu and Rosemarie H. Ziedonis, "Patents as Quality Signals for Entrepreneurial Ventures," *Academy of Management Proceedings* 2008, no. 1 (2008): 1–6, https://doi.org/10.5465/ambpp.2008.33653924.

9. Wesley M. Cohen, Richard R. Nelson, and John P. Walsh, *Protecting Their Intellectual Assets: Appropriability Conditions and Why U.S. Manufacturing Firms Patent (or Not)*, NBER Working Paper 7552 (National Bureau of Economic Research, 2000), http://doi.org/10.3386/w7552.

10. Hiral Patel, Laia Marin i Sola, Benjamin M. Theurer et al., *Cultured Meat: From Lab to Fork*, Barclays, November 18, 2021, https://fudzs.com/wp-content/uploads/2022/09/Cultured-Meat-From-Lab-to-Fork-Barclays.pdf (the survey covered China, India, Brazil, the United States, and the UK).

11. Bryant and Julie Barnett, "Consumer Acceptance of Cultured Meat: An Updated Review (2018–2020)," *Applied Sciences* 10, no. 15 (2020): 5201, https://doi.org/10.3390/app10155201.

12. "Questionnaire," *Consumer Food Insights* 3, no. 3 (March 2024), Purdue Center for Food Demand Analysis and Sustainability College of Agriculture, https://ag.purdue.edu/cfdas/wp-content/uploads/2024/04/Questionnaire_202403.pdf.

13. *Consumer Food Insights* 3, no. 3 (March 2024), Purdue Center for Food Demand Analysis and Sustainability College of Agriculture, https://ag.purdue.edu/cfdas/wp-content/uploads/2024/04/Report_202403-3.pdf.

14. Bella Nichole Kantor and Jonathan Kantor, "Public Attitudes and Willingness to Pay for Cultured Meat: A Cross-Sectional Experimental Study," Frontiers in Sustainable Food Systems 5 (2021), https://doi.org/10.3389/fsufs.2021.594650.

15. Mark Chong, Angela Leung, and Tricia Marjorie Fernandez, "On-Site Sensory Experience Boosts Acceptance of Cultivated Chicken," *Future Food* 9 (2024): 100301, https://doi.org/10.1016/j.fufo.2024.100326.

16. Bryant and Barnett, "Consumer Acceptance of Cultured Meat," https://doi.org/10.3390/app10155201.

17. "Condemnation of Poultry Carcasses Affected With Any Form of Avian Leukosis Complex; Recission," 88 Federal Register 55909, August 17, 2023, https://www.federalregister.gov/d/2023-17451.

18. 9 C.F.R. § 311.11 (2025).

19. Ronald O. Bates and Barbara Straw, "Hernias in Growing Pigs," Michigan State University *Pork Quarterly* 13 (2008), https://www.thepigsite.com/articles/hernias-in-growing-pigs.

20. Rachel Rabkin Peachman, "The Quest for Safer Chicken," *Consumer Reports*, August 4, 2021, https://www.consumerreports.org/chicken/the-quest-for-safer-chicken-food-safety/.

21. "Chicken and Food Poisoning," US Centers for Disease Control and Prevention, April 29, 2024, https://www.cdc.gov/food-safety/foods/chicken.html.

22. 21 C.F.R. Part 556 (2025).

23. Dhruv Kishore Bol, "AMR Threats from Meat and Meat Products," *Food Safety Magazine*, November 7, 2022, https://www.food-safety.com/articles/8121-amr-threats-from-meat-and-meat-products; Moyosore Joseph Adegbeye, Babatunde Oluwafemi Adetuyi, Anem I. Igirigi, Abosede Adisa, Valiollah Palangi, Susanna Aiyedun et al., "Comprehensive Insights Into Antibiotic Residues in Livestock Products: Distribution, Factors, Challenges, Opportunities, and Implications for Food Safety and Public Health," *Food Control* 163 (2024): 110545, https://doi.org/10.1016/j.foodcont.2024.110545.

Notes

24. "Drug Residue Warning Letters," US Food & Drug Administration, last updated July 13, 2023, https://www.fda.gov/animal-veterinary/compliance-enforcement/drug-residue -warning-letters.
25. "After Sampling Shows Drugs in 20 Percent of 'Antibiotic-Free' Meat, USDA to Crack Down on Label Claims," *Food Safety Magazine*, August 29, 2024, https://www.food-safety .com/articles/9706-after-sampling-shows-drugs-in-20-percent-of-antibiotic-free-meat -usda-to-crack-down-on-label-claims.
26. Nicholas Kristof, "Arsenic in Our Chicken?," *New York Times*, April 4, 2012, https://www .nytimes.com/2012/04/05/opinion/kristof-arsenic-in-our-chicken.html.
27. Naba Kumar Mondal, "Prevalence of Arsenic in Chicken Feed and Its Contamination Pattern in Different Parts of Chicken Flesh: A Market Basket Study," Environmental Monitoring and Assessment 192, no. 9 (2020): 590, https://doi.org/10.1007/s10661-020-08558-x.
28. Elsie M. Sunderland, Miling Li, and Kurt Bullard, "Decadal Changes in the Edible Supply of Seafood and Methylmercury Exposure in the United States," Environmental Health Perspectives 126, no. 1 (2018), https://doi.org/10.1289/EHP2644.
29. Mercury and Tuna: U.S. Advice Leaves Lots of Questions Peter Waldman, "Mercury and Tuna: U.S. Advice Leaves Lots of Questions," *Wall Street Journal*, August 1, 2005, https:// www.wsj.com/articles/SB112268169016100484.
30. Tabuchi, "Why Is Mercury Stubbornly High in Tuna? Researchers Might Have an Answer," *New York Times*, February 27, 2024, https://www.nytimes.com/2024/02/27/climate /tuna-mercury.html.
31. Leonardo Trasande, Philip J. Landrigan, and Clyde Schechter, "Public health and economic consequences of methyl mercury toxicity to the developing brain," *Environmental Health Perspectives* 113, no. 5 (2005): 590–596, https://doi.org/10.1289/ehp.7743.

Chapter 7

1. Richard Branson, "Tasting the Clean Meat of the Future," Virgin Airlines, October 31, 2019, https://www.virgin.com/branson-family/richard-branson-blog/tasting-clean-meat-future.
2. Nick Rufford, "Can the Meat-Free Impossible Burger Save the World?," *Sunday Times Driving*, April 19, 2017, https://www.driving.co.uk/news/lifestyle/impossible-burger/.
3. "Social and Environmental Impacts of Alternative Proteins," McKinsey and Conservation International, October 2022 (on file with author), slide 36 (P/T/N: 33/31/15); Chris Bryant, Mathilde Alexandre, Elsa Guadarrama, Indy Kaur, and Gill Hayes, "What We Know About UK Plant-Based Meat Consumers," Bryant Research, October 2023, https:// bryantresearch.co.uk/insight-items/uk-pbm-consumers/; *Power of Plant-Based Food & Beverages* (FMI and NielsenIQ, 2022), https://www.fmi.org/forms/store/ProductFormPublic /power-of-plant-based-foods.
4. Survey from IRI Melissa Sue Sorrells, "Barriers to Alt-Meat Adoption," Alt-Meat, December 28, 2022, https://www.alt-meat.net/barriers-alt-meat-adoption.
5. Peter Slade, "If You Build It, Will They Eat It? Consumer Preferences for Plant-Based and Cultured Meat Burgers," *Appetite* 125, no. 1 (2018): 428–437, https://doi.org/10.1016/j .appet.2018.02.030 (32% of consumers would choose either plant-based or cultivated meat: 21% plant-based, 11% cultivated); Glynn T. Tonsor, Jayson L. Lusk, and Ted C. Schroeder, *Impact of New Plant-Based Protein Alternatives on U.S. Beef Demand* (Prepared for the Cattlemen's Beef Promotion and Research Board, January 17, 2021) https://www.agmanager.info /livestock-meat/meat-demand/meat-demand-research-studies/impact-new-plant-based

Notes

-protein-0 (more than one-quarter would choose plant-based meat); Lewis Bollard, "If We Build It, Will They Come?," Open Philanthropy, March 9, 2021, https://www .openphilanthropy.org/research/if-we-build-it-will-they-come/.

6. "Insight of the Week: Significant Opportunities for Plant-Based Alternatives to Real Meat, Especially in Asia and Latin America," Globe Scan, February 26, 2021.

7. Slade, "If You Build It, Will They Eat It?"

8. William J Bulsiewicz, "The Importance of Dietary Fiber for Metabolic Health" *American Journal of Lifestyle Medicine* 17, no. 5 (2023): 639-648, https://doi.org/10.1177/15598276231167778.

9. "Plant-Based Meat Nutrition: The Facts," Good Food Institute, August 2022, https://gfi.org /wp-content/uploads/2022/11/Plant-based-meat-nutrition-fact-sheet.pdf.

10. Bruce Friedrich, "TASTY Award Winners Nutrition Summary," May 28, 2025, https://docs .google.com/spreadsheets/d/15_CUx7ypY71wncKw8kt70d1gQmy3BeDQh1IrjDgJMSs /edit (on file with author).

11. Kevin D. Hall, Alexis Ayuketah, Robert Brychta, Hongyi Cai, Thomas Cassimatis, Kong Y. Chen et al., "Ultra-Processed Diets Cause Excess Calorie Intake and Weight Gain: An Inpatient Randomized Controlled Trial of Ad Libitum Food Intake," *Cell Metabolism* 30, no. 1 (2019): 67-77.e3, https://doi.org/10.1016/j.cmet.2019.05.008.

12. Andrea Petersen, "The Trouble With America's Ultra-Processed Diet," *Wall Street Journal*, November 14, 2023, https://www.wsj.com/health/wellness/ultra-processed-foods-dietary -guidelines-de00ccaa.

13. Artemis P. Simopoulos, "The Importance of the Ratio of Omega-6/Omega-3 Essential Fatty Acids," *Biomedicine & Pharmacotherapy* 56, no. 8 (2002): 365–379, https://doi.org/10.1016 /s0753-3322(02)00253-6 ("Excessive amounts of omega-6 polyunsaturated fatty acids (PUFA) and a very high omega-6/omega-3 ratio, as is found in today's Western diets, promote the pathogenesis of many diseases, including cardiovascular disease, cancer, and inflamma-tory and autoimmune diseases . . . In the secondary prevention of cardiovascular disease, a ra-tio of 4/1 was associated with a 70% decrease in total mortality"); Yuchen Zhang, Yitang Sun, Qi Yu, Suhang Song, J. Thomas Brenna, Ye Shen et al., "Higher Ratio of Plasma Omega-6/ Omega-3 Fatty Acids Is Associated With Greater Risk of All-Cause, Cancer, and Cardio-vascular Mortality: A Population-Based Cohort Study in UK Biobank," *eLife* (2004), https:// doi.org/10.7554/eLife.90132.3.

14. Melissa M. Lane, Elizabeth Gamage, Shutong Du, Deborah N. Ashtree, Amelia J. McGuin-ness, Sarah Gauci et al., "Ultra-Processed Food Exposure and Adverse Health Outcomes: Umbrella Review of Epidemiological Meta-Analyses," *BMJ* 384 (2024): e077310, https://doi .org/10.1136/bmj-2023-077310.

15. Reynalda Cordova, Vivian Viallon, Emma Fontvieille, Laia Peruchet-Noray, Anna Jansana, Karl-Heinz Wagner et al., "Consumption of Ultra-Processed Foods and Risk of Multimor-bidity of Cancer and Cardiometabolic Diseases: A Multinational Cohort Study," *Lancet Regional Health Europe* 35 (2023): 100771, https://doi.org/10.1016/j.lanepe.2023.100771; Francesco Visioli, Daniele Del Rio, Vincenzo Foglianoc, Franca Marangonid, and Andrea Poli, "Ultra Processed Foods and Cancer," *Lancet Regional Health Europe* 38 (2024): 100863, https://doi.org/10.1016/j.lanepe.2024.100863 ("the association between UPF consumption and the risk of multimorbidity would disappear if the data were adjusted not only for the consumption of sugary or artificially sweetened beverages, but also for foods of animal origin at the same time . . . [T]he article underlines the absolute need to return to the evaluation of

Notes

foods on the basis of their nutritional role (including their nutrient composition, quantities consumed, metabolic effects, etc.) and not on the basis of their degree of processing").

16. Samuel J. Dicken, Christina C. Dahm, Daniel B. Ibsen, Anja Olsen, Anne Tjønneland, Mariem Louati-Hajji et al., "Food consumption by degree of food processing and risk of type 2 diabetes mellitus: a prospective cohort analysis of the European Prospective Investigation into Cancer and Nutrition (Epic)." *Lancet Regional Health Europe.* 2024;46:101043.

17. Fernanda Rauber, Maria Laura da Costa Louzada, Kiara Chang, Inge Huybrechts, Marc J. Guntere, Carlos Augusto Monteiro et al., "Implications of Food Ultra-Processing on Cardiovascular Risk Considering Plant Origin Foods: An Analysis of the UK Biobank Cohort," *Lancet Regional Health Europe* 43 (2024): 100948, https://doi.org/10.1016/j.lanepe.2024.100948. See Appendix A for examples of food categories, including meat alternatives.

18. Youmshajekian, "Are Plant-Based Ultraprocessed Foods Linked to Heart Disease?," *Scientific American*, June 24, 2024, https://www.scientificamerican.com/article/what-a-study -on-ultraprocessed-fake-meat-and-heart-disease-really-found/.

19. Anthony Crimarco, Sparkle Springfield, Christina Petlura, Taylor Streaty, Kristen Cunanan, Justin Lee et al., "A Randomized Crossover Trial on the Effect of Plant-Based Compared With Animal-Based Meat on Trimethylamine-N-Oxide and Cardiovascular Disease Risk Factors in Generally Healthy Adults: Study With Appetizing Plantfood—Meat Eating Alternative Trial (SWAP-MEAT)," *American Journal of Clinical Nutrition* 112, no. 5 (2020): 1188–1199. https://doi.org/10.1093/ajcn/nqaa203.

20. Matthew Nagra, Felicia Tsam, Shaun Ward, Ehud Ur, "Animal vs Plant-Based Meat: A Hearty Debate," *Canadian Journal of Cardiology* 40, no. 7 (2024): 1198–1209, https://doi .org/10.1016/j.cjca.2023.11.005; Rubén Fernández-Rodríguez, Bruno Bizzozero-Peroni, Valentina Díaz-Goñi, Miriam Garrido-Miguel, Gabriele Bertotti, Alberto Roldán-Ruiz et al., "Plant-Based Meat Alternatives and Cardiometabolic Health: A Systematic Review and Meta-Analysis," *American Journal of Clinical Nutrition* 121, no. 2 (2025): 274–283, https:// doi.org/10.1016/j.ajcnut.2024.12.002; Christopher J. Bryant, "Plant-Based Animal Product Alternatives Are Healthier and More Environmentally Sustainable Than Animal Products," *Future Foods* 6 (2022): 100174, https://doi.org/10.1016/j.fufo.2022.100174.

21. Karen Formanski, *2021 State of the Industry Report: Plant-Based Meat, Seafood, Eggs, and Dairy*, Good Food Institute, 2022, https://gfi.org/wp-content/uploads/2022/04/2021-Plant -Based-State-of-the-Industry-Report-1.pdf, 83; Ryan Dowdy, "Plant-Based Meat: Anticipating 2030 Production Requirements," Good Food Institute, 2022, https://gfi.org/wp -content/uploads/2022/01/Production-volume-modeling-deck-v3.pdf, slide 4 (sources on file with author); *Animal Meat Disappears From the Shelves New Alternatives Are Whirling the Agricultural and Food Industries Upside Down* (Kearney, May 20, 2019).

22. *The Global Food System: Identifying Sustainable Solutions* (Credit Suisse, June 2021), 60, figure 28, https://www.foodfrontier.org/wp-content/uploads/2021/06/Credit-Suisse-The-global -food-system-identifying-sustainable-solutions.pdf.

23. Tad Friend, "Can a Burger Help Solve Climate Change," *New Yorker*, September 23, 2019, https:// www.newyorker.com/magazine/2019/09/30/can-a-burger-help-solve-climate-change.

24. Vincenzina Caputo, Giovanni Sogari, and Ellen J. Van Loo, "Do Plant-Based and Blend Meat Alternatives Taste Like Meat? A Combined Sensory and Choice Experiment Study," *Applied Economic Perspectives and Policy* 45, no. 1 (2022): 86–105, https://doi.org/10.1002 /aepp.13247.

Notes

25. Lewis Bollard, "How to Lower the Price of Plant-Based Meat," Open Philanthropy, August 12, 2020, https://www.openphilanthropy.org/research/how-to-lower-the-price-of-plant -based-meat/.

26. Steven Finn, "Why Meati Failed: A Lesson in Tech and Consumer Brands," LinkedIn, May 7, 2025, https://www.linkedin.com/posts/steven-finn-bb206011_many-people-in-the-food -tech-ecosystem-have-activity-7325974047974957056-91KV/.

27. *Taste of the Industry 2025* (NECTAR, March 2025), https://www.nectar.org/sensory -research/2025-taste-of-the-industry.

28. Battle et al. *2024 State of the Industry.*

29. "Foodtech Project Call on Alternative Proteins," *Enterprise Singapore*, accessed June 15, 2025, https://www.enterprisesg.gov.sg/grow-your-business/innovate-with-us/market-access-and -networks/global-innovation-alliance/cip/foodtech-project-call-on-alternative-proteins.

30. *Future of the Industry 2024* (NECTAR, September 2024), https://static1.squarespace.com /static/66107b8b2226bc6925086385/t/67d8aa880b93676eec726805/1742252681238/Nectar _FOTI+2024_General+Report+Summary+Balanced+Protein+One-Pager.pdf.

Chapter 8

1. David E. Sanger, "Home Computer Is Out in the Cold," *New York Times*, February 20, 1985, https://www.nytimes.com/1985/02/20/garden/home-computer-is-out-in-the-cold.html.

2. Walter Isaacson, *Steve Jobs* (Simon & Schuster, 2011) (from the conclusion of the biography authored by Isaacson).

3. David Brooks, "'Human Beings Are Soon Going to Be Eclipsed'," *New York Times*, July 13, 2023, https://www.nytimes.com/2023/07/13/opinion/ai-chatgpt-consciousness-hofstadter .html; Ajeya Cotra, "Language Models Surprised Us," Planned Obsolescence, August 29, 2023, https://www.planned-obsolescence.org/language-models-surprised-us/.

4. Dean Bubley, "Dean Bubley's Disruptive Wireless: Thought-Leading Wireless Industry Analysis," Disruptive Analysis, May 2006, https://disruptivewireless.blogspot.com/2006/05 /device-divergence-always-beats.html. ("'The MP3 player is dead' 'Standalone digital cameras on the decline.' A single converged device for everything! Yeah right.")

5. "In the City," part of the *Horse* exhibition, American Museum of Natural History, accessed June 15, 2025, https://www.amnh.org/exhibitions/horse/how-we-shaped-horses-how-horses -shaped-us/work/in-the-city (reports that horses excrete 45 pounds of manure/day and two gallons of urine, but most other sources have lower numbers); "Horse-Drawn Vehicles in the City," The Henry Ford, September 13, 2021, https://www.thehenryford.org/explore/blog /horse-drawn-vehicles-in-the-city.

6. Águeda García de Durango, "New York, Manure and Stairs: When Horses Were the Cities' Nightmares," *Smart Water Magazine*, June 9, 2019, https://smartwatermagazine.com /blogs/agueda-garcia-de-durango/new-york-manure-and-stairs-when-horses-were-cities -nightmares; Diana Wolf, "The Surprising Source of New York's First Environmental Problem," *Mental Floss*, June 5, 2008, https://www.mentalfloss.com/article/18783/surprising -source-new-yorks-first-environmental-problem; Ben Johnson, "The Great Horse Manure Crisis of 1894," Historic UK, accessed June 15, 2025, https://www.historic-uk.com /HistoryUK/HistoryofBritain/Great-Horse-Manure-Crisis-of-1894. While you're here in the endnotes, I'll also mention that there's no record of Henry Ford saying that if he'd asked his customers what they wanted, they would have said a faster horse.

7. "Horse-Drawn Vehicles in the City," The Henry Ford; "In the City," American Museum of

Notes

Natural History; Jennifer Lee, "When Horses Posed a Public Health Hazard," *New York Times*, June 9, 2008, https://archive.nytimes.com/cityroom.blogs.nytimes.com/2008/06/09 /when-horses-posed-a-public-health-hazard/; Johnson, "The Great Horse Manure Crisis of 1894"; David Doochin, "The First Global Urban Planning Conference Was Mostly About Manure," *Atlas Obscura*, July 29, 2016, https://www.atlasobscura.com/articles /the-first-global-urban-planning-conference-was-mostly-about-manure.

8. Alexander Winton, "Get A Horse! America's Skepticism Toward the First Automobiles," *Saturday Evening Post*, February 8, 1930, https://www.saturdayeveningpost.com/2017/01 /get-horse-americas-skepticism-toward-first-automobiles/.

9. "What it Was Like to Drive Over 100 Years Ago," Edgar Snyder & Associates, accessed June 15, 2025, https://www.edgarsnyder.com/resources/what-it-was-like-to-drive-100-years-ago.

10. Steven Klepper, "The Evolution of the U.S. Automobile Industry and Detroit as Its Capital," November 2001, https://ftp.zew.de/pub/zew-docs/entrepreneurship/klepper.pdf.

11. "1920-1930," Cars Amber & Madison, accessed June 15, 2025, https://transportationchanges. weebly.com/1920-1930.html.

12. Jim Rasenberger, "1908: Aeroplanes! Skyscrapers! The Race to the North Pole! Mobile Phones? Inventions, Predictions and Breakthroughs That Propelled America Into the Modern Age," *Smithsonian Magazine*, January 2008, https://www.smithsonianmag.com /history/1908-7683115/.

13. Ed Ballard, "Now for Some Good News About Climate: Costs for Renewables Have Plummeted and Growth Is Exceeding Expectations," *Wall Street Journal*, November 27, 2023, https://www.wsj.com/business/energy-oil/now-for-some-good-news-about-climate-27236f56; Michael Liebreich, "In Energy and Transportation, Stick It to the Orthodoxy!," *Bloomberg New Energy Finance*, September 27, 2017, https://about.bnef.com/blog/energy -transportation-stick-orthodoxy/ (IEA's optimistic model was for 50 gigawatt/year increases through 2040).

14. "Sun, Wind and Drain," *Economist*, July 29, 2014, https://www.economist.com/finance -and-economics/2014/07/29/sun-wind-and-drain.

15. "Total Renewable Capacity Additions by Technology, 2019-2024," International Energy Agency, last updated March 23, 2025, https://www.iea.org/data-and-statistics/charts /total-renewable-capacity-additions-by-technology-2019-2024 (553 GW in 2024).

16. "Dawn of the Solar Age," *Economist*, June 22, 2024, https://www.economist.com /weeklyedition/2024-06-22.

17. *World Energy Outlook 2015* (International Energy Agency, 2015), https://iea.blob.core .windows.net/assets/5a314029-69c2-42a9-98ac-d1c5deeb59b3/WEO2015.pdf.

18. Charles Lane, "The Electric Car Mistake," *Washington Post*, February 11, 2013, https://www.washingtonpost.com/opinions/charles-lane-obamas-electric-car-mistake /2013/02/11/441b39f6-7490-11e2-aa12-e6cf1d31106b_story.html.

19. *X-Change: Batteries The Battery Domino Effect* (RMI, December 2023), https://rmi.org /wp-content/uploads/dlm_uploads/2023/12/xchange_batteries_the_battery_domino_effect .pdf, 17.

20. Zhiwei Guo, Tao Li, Bowen Shi, and Hongchao Zhang, "Economic Impacts and Carbon Emissions of Electric Vehicles Roll-Out Towards 2025 Goal of China: An Integrated Input-Output and Computable General Equilibrium Study," *Sustainable Production and Consumption* 31 (2022): 165–174, https://doi.org/10.1016/j.spc.2022.02.009.

21. Eric Beinhocker and J. Doyne Farmer, "The Clean Energy Revolution Is

Notes

Unstoppable," *Wall Street Journal*, February 28, 2025, https://www.wsj.com/business/energy-oil/thecleanenergyrevolution-is-unstoppable-88af7ed5; see also Kingsmill Bond, Sam Butler-Sloss, and Daan Walter, *The Cleantech Revolution It's Exponential, Disruptive, and Now* (RMI, 2024), https://rmi.org/insight/the-cleantech-revolution/.

22. Rupert Way, Matthew Ives, Penny Mealy, and J. Doyne Farmer, "Empirically Grounded Technology Forecasts and the Energy Transition," *Joule* 6, no. 9 (2022): 2057–2082, https://doi.org/10.1016/j.joule.2022.08.009; Beinhocker and Farmer, "The Clean Energy Revolution Is Unstoppable."

23. Teo Lombardo, Leonardo Paoli, Araceli Fernandez Pales, and Timur Gül, "The Battery Industry Has Entered a New Phase," International Energy Agency, March 5, 2025, https://www.iea.org/commentaries/the-battery-industry-has-entered-a-new-phase.

24. "Lower Battery Prices Are Expected to Eventually Boost EV Demand," Goldman Sachs, February 29, 2024, https://www.goldmansachs.com/insights/articles/even-as-ev-sales-slow-lower-battery-prices-expect.

25. Roser, "Learning Curves: What Does It Mean for a Technology to Follow Wright's Law?," Our World in Data, April 18, 2023, https://ourworldindata.org/learning-curve; Béla Nagy, J. Doyne Farmer, Quan M. Bui, Jessika E. Trancik, "*Statistical Basis for Predicting Technological Progress*," PLoS ONE 8, no. 2 (2013): e52669, https://doi.org/10.1371/journal.pone.0052669.

26. Femke J. M. M. Nijsse, Jean-Francois Mercure, Nadia Ameli, Francesca Larosa, Sumit Kothari, Jamie Rickman et al., "The Momentum of the Solar Energy Transition," *Nature Communications* 14, article no. 6542 (2023), https://doi.org/10.1038/s41467-023-41971-7; Max Roser, "Learning Curves: What Does It Mean for a Technology to Follow Wright's Law?," Our World in Data, April 18, 2023, https://ourworldindata.org/learning-curve.

27. Leslie Roberts, "Controversial From the Start," *Science* 291, no. 5507 (2001): 1182–1188, https://doi.org/10.1126/science.291.5507.1182a; see also Zachary M. Utz, "'The Human Genome Project Is Simply a Bad Idea,'" National Human Genome Research Institute, last updated May 6, 2024, https://www.genome.gov/virtual-exhibits/human-genome-project-is-simply-a-bad-idea (discussing a collection of 55 letters from scientists at 33 academic institutions opposing the project).

28. Maynard V. Olson, "A Behind-the-Scenes Story of Precision Medicine," *Genomics Proteomics Bioinformatics* 15, no. 1 (20173-10), https://doi.org/10.1016/j.gpb.2017.01.002; John Hodgson, "Gene Sequencing's Industrial Revolution," IEEE Spectrum, November 1, 2000, https://spectrum.ieee.org/gene-sequencings-industrial-revolution.

29. Roberts, "Controversial From the Start."

30. Razib Khan, "Genomics Has Revealed An Age Undreamed Of," Palladium, November 17, 2023, https://www.palladiummag.com/2023/11/17/genomics-has-revealed-an-age-undreamed-of/.

31. Hodgson, "Gene Sequencing's Industrial Revolution."

32. Khan, "Genomics Has Revealed An Age Undreamed Of."

33. Chunka Mui, Paul B. Carroll, and Tim Andrews, *A Brief History of a Perfect Future: Inventing the world we can proudly leave our kids by 2050* (Future Histories Press, 2021).

34. The History of Ice & the IPIA, International Packaged Ice Association, accessed June 15, 2025, https://www.packagedice.com/history-of-ice.html.

35. Louis Anslow, "Aeroplanes, Airships and Beta Bias," Pessimists Archive, February 8, 2024, https://newsletter.pessimistsarchive.org/p/aeroplanes-airships-and-beta-bias.

36. Louis Anslow, "Why Steve Wozniak Dismissed PCs and Steve Jobs Didn't," Pessimists Archive, May 6, 2024, https://newsletter.pessimistsarchive.org/p/why-steve-wozniak-dismissed

Notes

-pcs-and; James Clive-Matthews, "A Brief History of Tech Skepticism," *Strategy+Business*, May 5, 2023, https://www.strategy-business.com/article/A-brief-history-of-tech-skepticism.

Chapter 9

1. Zane Swanson, Caitlin Welsh, and Joseph Majkut, *Mitigating Risk and Capturing Opportunity: The Future of Alternative Proteins*, Center for Strategic & International Studies, May 11, 2023, https://www.csis.org/analysis/mitigating-risk-and-capturing-opportunity-future-alternative-proteins.

2. Alexander Fleming, "Penicillin," *British Medical Journal* 2, no. 4210 (1941): 386, https://www.bmj.com/content/2/4210/386; Robert Gaynes, "The Discovery of Penicillin—New Insights After More Than 75 Years of Clinical Use," *Emerging Infectious Diseases* 23, no. 5 (2017): 849–853, https://doi.org/10.3201/eid2305.161556.

3. "Discovery and Development of Penicillin," American Chemistry Society, accessed June 15, 2025, https://www.acs.org/education/whatischemistry/landmarks/flemingpenicillin.html.

4. Gary Clyde Hufbauer and Euijin Jung, *Scoring 50 Years of US Industrial Policy, 1970–2020*, PIIE Briefing 21-5 (Peterson Institute for International Economics, November 2021), https://www.piie.com/sites/default/files/documents/piieb21-5.pdf.

5. Richard Nixon, "Address on the State of the Union Delivered Before a Joint Session of the Congress," (speech, Washington, D.C., January 30, 1974), American Presidency Project, https://www.presidency.ucsb.edu/node/256218.

6. "Energy Technology RD&D Budgets Data Explorer," International Energy Agency, last updated May 15, 2025, https://www.iea.org/data-and-statistics/data-tools/energy-technology-rdd-budgets-data-explorer.

7. Eric Beinhocker and J. Doyne Farmer, "The Clean Energy Revolution Is Unstoppable," *Wall Street Journal*, February 28, 2025, https://www.wsj.com/business/energy-oil/thecleanenergyrevolution-is-unstoppable-88af7ed5; "Solar (Photovoltaic) Panel Prices vs. Cumulative Capacity," Our World in Data, last updated November 15, 2024, https://ourworldindata.org/grapher/solar-pv-prices-vs-cumulative-capacity?time=1998..latest.

8. "Share of Electricity Generated by Solar Power," Our World in Data, last updated May 12, 2025, https://ourworldindata.org/grapher/share-electricity-solar?tab=chart&country=OWID_WRL~OWID_EU27~DEU~JPN~USA~CHN (the US hit .01% in 2007, 1% in 2015, and 6.91% in 2024; China passed .01% in 2010, 1% in 2016, and 8.28% in 2024; the world: 2002 & 2015 for .01% and 1%; 6.91% in 2024).

9. Taalbi and Hana Nielsen, "The Role of Energy Infrastructure in Shaping Early Adoption of Electric and Gasoline Cars," *Nature Energy* 6 (2021): 970–976, https://doi.org/10.1038/s41560-021-00898-3; "A new study argues that insufficient infrastructure doomed the first electric cars," *Economist*, October 30, 2021, https://www.economist.com/graphic-detail/2021/10/30/a-new-study-argues-that-insufficient-infrastructure-doomed-the-first-electric-cars.

10. "The History of the Electric Car," US Department of Energy, September 15, 2014, https://www.energy.gov/articles/history-electric-car.

11. Linda Lew, "Electric Vehicle Battery Packs See Biggest Price Drop Since 2017," Bloomberg, December 10, 2024, https://www.bloomberg.com/news/articles/2024-12-10/electric-vehicle-battery-packs-see-biggest-price-drop-since-2017.

12. Sutton et al., "Recipe for a Livable Planet."

13. *U.S. Pharmaceutical Market Size, Share & Trends Analysis Report By Molecule (Biologics & Biosimilars, Conventional Drugs), By Product, By Type, By Route Of Administration, By Disease,*

Notes

By Age Group, By Distribution Channel, And Segment Forecasts, 2025 - 2030, Report GVR-4-68040-210-9 (Grand View Research, 2024), https://www.grandviewresearch.com/industry-analysis/us-pharmaceuticals-market-report.

14. "About NIH," US National Institutes of Health, last reviewed March 13, 2025, https://www.nih.gov/about-nih.

15. "Budget," US National Institutes of Health, last reviewed June 13, 2025, https://www.nih.gov/about-nih/what-we-do/budget.

16. Ekaterina Galkina Cleary, Matthew J. Jackson, Edward W. Zhou, and Fred D. Ledley, "Comparison of Research Spending on New Drug Approvals by the National Institutes of Health vs the Pharmaceutical Industry, 2010-2019," *JAMA Health Forum* 4, no. 4 (2023): e230511, https://doi.org/10.1001/jamahealthforum.2023.0511 ("The products without NIH funding were a chelating agent and osmotic laxative").

17. Colin Gordon, "We Don't Have to Keep Shoveling Money to Big Ag," Jacobin, January 9, 2021, https://jacobin.com/2021/01/big-ag-farm-subsidies-agriculture.

18. Chris Edwards, "Cutting Federal Farm Subsidies," Briefing Paper No. 162, Cato Institute, August 31, 2023, https://www.cato.org/briefing-paper/cutting-federal-farm-subsidies.

19. Kelly P. Nelson and Keith Fuglie, "Investment in U.S. Public Agricultural Research and Development Has Fallen by a Third Over Past Two Decades, Lags Major Trade Competitors," US Department of Agriculture, Economic Research Service, June 6, 2022, https://www.ers.usda.gov/amber-waves/2022/june/investment-in-u-s-public-agricultural-research-and-development-has-fallen-by-a-third-over-past-two-decades-lags-major-trade-competitors; Kelly P. Nelson, "Agricultural and Food Research and Development Expenditures in the United States," US Department of Agriculture, Economic Research Service, last updated May 12, 2025, https://www.ers.usda.gov/data-products/agricultural-and-food-research-and-development-expenditures-in-the-united-states/.

20. "CAP Expenditure," European Commission, accessed April 4, 2025, https://agriculture.ec.europa.eu/data-and-analysis/financing/cap-expenditure_en.

21. Anniek J. Kortleve, José M. Mogollón, Helen Harwatt, and Paul Behrens, "Over 80% of the European Union's Common Agricultural Policy Supports Emissions-Intensive Animal Products," *Nature Food* 5 (2024): 288–292, https://doi.org/10.1038/s43016-024-00949-4.

22. "Forbes' World Billionaires List: The Richest in 2025," *Forbes*, March 15, 2025, https://www.forbes.com/billionaires/ (Elon Musk, Mark Zuckerberg, Jeff Bezos, Larry Ellison, Warren Buffet, Larry Page, Sergey Brin, and Steve Ballmer).

23. National Research Council, *Funding a Revolution: Government Support for Computing Research* (National Academies of Sciences, Engineering, and Medicine, 1999), https://doi.org/10.17226/6323; Shane Greenstein, *How the Internet Became Commercial: Innovation, Privatization, and the Birth of a New Network* (Princeton University Press, 2015); Mariana Mazzucato, *The Entrepreneurial State: Debunking Public vs. Private Sector Myths* (Anthem Press, 2013).

24. Research Council, *Funding a Revolution*; Greenstein, *How the Internet Became Commercial*; Mazzucato, *The Entrepreneurial State*.

25. Thomas Haigh, "There Was No 'First AI Winter'," Communications of the ACM 66, no. 12 (2023): 35–39, https://doi.org/10.1145/3625833.

26. "Center for Strategic and International Studies – Bias and Credibility," Media Bias/Fact Check, accessed June 15, 2025, https://mediabiasfactcheck.com/center-for-strategic-and-international-studies/; "Public Policy Research Think Tanks: Top Think Tanks—US," Penn Libraries Guide, University of Pennsylvania, https://guides.library.upenn.edu/c

Notes

.php?g=1035991&p=7509974; James G. McGann, *2020 Global Go To Think Tank Index Report*, TTCSP Global Go To Think Tank Index Reports 18 (Think Tanks and Civil Societies Program, University of Pennsylvania, 2021). https://repository.upenn.edu/think_tanks/18.

27. Although GFI sponsored the work, we had no control over the final product, which was led by CSIS Director and Deputy Director of Global Food and Water Security Caitlin Welsh and Zane Swanson, as well as CSIS Director of Energy Security and Climate Change Joseph Majkut. As with all CSIS program staff, the credentials of this team are exceptional: Welsh served under Presidents Obama and Trump at the State Department and then National Security Council, most recently as the NSC's Director for Global Economic Engagement. Swanson earned a PhD in Evolutionary Anthropology from Duke, where his work focused on food and water security among pastoralist communities in rural Kenya. Finally, Majkut earned his PhD from Princeton in atmospheric and oceanic sciences before working on climate policy for more than a decade, including as Director of Climate Policy at the Niskanen Center.

28. "Meat – Worldwide," Statistica, accessed June 15, 2025, https://www.statista.com/outlook /cmo/food/meat/worldwide (land animal meat at $1.55 trillion in 2025); *The State of World Fisheries and Aquaculture*, Food and Agriculture Organization of the United Nations, https:// openknowledge.fao.org/server/api/core/bitstreams/f985caed-cc7a-457e-8107-7ce16c6ef209 /content (seafood market valued at $472 billion in 2022).

29. "Digital Report Launch: The Future Appetite for Alternative Proteins," Center for Strategic and International Studies (CSIS), May 4, 2023, https://www.csis.org/analysis/digital -report-launch-future-appetite-alternative-proteins.

30. *Charting the Future of Biotechnology* (National Security Commission on Emerging Biotechnology, April 2025), https://www.biotech.senate.gov/wp-content/uploads/2025/04/NSCEB -Full-Report-%E2%80%93-Digital-%E2%80%934.28.pdf.

31. Seung Yun Lee, Da Young Lee, Seung Hyeon Yun, Juhyun Lee, Ermie Mariano Jr., Jinmo Park et al., "Current Technology and Industrialization Status of Cell-Cultivated Meat," *Journal of Animal Science and Technology* 66, no. 1 (2024): 1–30, https://doi.org/10.5187/jast.2023.e107.

32. "ESA Investigates Cultured Meat as Novel Space Food," European Space Agency, June 1, 2021, https://www.esa.int/Enabling_Support/Preparing_for_the_Future/Discovery_and_Prepa- ration/ESA_investigates_cultured_meat_as_novel_space_food; "On the Road to Cultured Meat for Astronauts (And Earthlings)," European Space Agency, March 31, 2022, https:// www.esa.int/Enabling_Support/Preparing_for_the_Future/Discovery_and_Preparation /On_the_road_to_cultured_meat_for_astronauts_and_Earthlings.

33. Marie Morales, "SpaceX Set to Conduct Experiment of Artificial Meat; Sends 4-Member Private Crew to ISS for the Tests," *Science Times*, last updated April 6, 2022, https://www .sciencetimes.com/articles/36962/20220404/fake-steak-space-x-set-conduct-experiment -artificial-meat-sends.htm.

34. "USDA and FDA Announce a Formal Agreement to Regulate Cell-Cultured Food Products from Cell Lines of Livestock and Poultry," US Department of Agriculture, March 7, 2019, https://www.usda.gov/about-usda/news/press-releases/2019/03/07/usda-and-fda-announce -formal-agreement-regulate-cell-cultured-food-products-cell-lines-livestock-and.

35. Secretary Sonny Perdue, BIO Virtual Fireside Chat (Sept. 22, 2020), https://bit.ly/3toJOv3.

36. "GCR: Laying the Scientific and Engineering Foundation for Sustainable Cultivated Meat Production," Award Abstract # 2021132, US National Science Foundation, accessed June 15, 2025, https://www.nsf.gov/awardsearch/showAward?AWD_ID=2021132.

37. Emiliano Bellini, "Global Solar Manufacturing Sector Now at 50% Utilization Rate, Says

IEA," *PV Magazine*, May 7, 2024, https://www.pv-magazine.com/2024/05/07/global-solar
-manufacturing-sector-now-at-50-utilization-rate-says-iea; *Solar Energy Industry Report
– 2024 – It's Not Just Sunshine, It's the Future* (Crisp Idea Research, 2024), https://www
.crispidea.com/report/solar-energy-industry-report-2024/; Ilkhan Ozsevim, "Global Battery
Industry Enters New Phase as Demand Surges and Prices Drop Below Key Cost Threshold
for Electric Vehicles," Automotive Manufacturing Solutions, March 7, 2025, https://www
.automotivemanufacturingsolutions.com/ev-battery-production/battery-prices-break-ice
-cost-barrier-opening-up-major-ev-potential/46870.article.

38. *EV Battery Market Size, Share, Growth, Trends, & Analysis, 2035*, Report AT 7840 (Markets
& Markets, 2025), https://www.marketsandmarkets.com/Market-Reports/electric-vehi-
cle-battery-market-100188347.html.

39. "China," International Energy Agency, accessed June 15, 2025, https://www.iea.org
/countries/china/energy-mix.

40. "Digital Report Launch," CSIS.

41. "Garbarino, Newhouse Push for U.S. Strategy to Protect U.S. Food Supply Chains and
Counter China's Advances in Biotechnology," (press release, September 30, 2024, Andrew R.
Garbarino (R-NY-02), Dan Newhouse (R-WA-04), Ken Calvert (R-CA-41), Lori Chavez-
DeRemer (R-OR-05), Brian Fitzpatrick (R-PA-01), Jennifer Kiggans (R-VA-02), Mike
Lawler (R-NY-17), Nancy Mace (R-SC-01), Marc Molinaro (R-NY-19), Joe Wilson (R-
SC-02), and Robert Wittman (R-VA-01)), https://garbarino.house.gov/media/press-releases
/garbarino-newhouse-push-us-strategy-protect-us-food-supply-chains-and-counter.

42. Bruce Friedrich, "The Next Global Agricultural Revolution," TED, April 2019, 5 min., 39
sec., https://www.ted.com/talks/bruce_friedrich_the_next_global_agricultural_revolution.

43. John R. Tyson, Jay Collins, Isobel Coleman, Liz Schrayer, and Chris Cleverly, "The Fight
for Food: Value Chains and Partnerships," 2022 Concordia Annual Summit, September 21,
2022, https://youtu.be/CYrctRjcpmg.

44. "Brazil's JBS Agrees to Buy Spanish Lab Meat Firm in $100 Million Push Into Sector," *Re-
uters*, November 17, 2021, https://www.reuters.com/article/world/europe/brazils-jbs-agrees
-to-buy-spanish-lab-meat-firm-in-100-million-push-into-secto-idUSKBN2I22HC/.

45. Emiko Terazono and Judith Evans, "Has the Appetite for Plant-Based Meat Already
Peaked?," *Financial Times*, January 26, 2022, https://www.ft.com/content/996330d5-5ffc
-4f35-b5f8-a18848433966.

46. "Alternative Protein Trends: 5 Things to Watch in 2024," Cargill, February 19, 2024, https://
www.cargill.com/story/alternative-protein-trends.

47. Joshua Poole, "Alt-Protein, Agriculture and AI: Cargill's CTO Unravels Macrotrends
in Human and Animal Nutrition," Food Ingredients First, February 26, 2024, https://
www.foodingredientsfirst.com/news/alt-protein-agriculture-and-ai-cargills-cto-unravels
-macrotrends-in-human-and-animal-nutrition.html.

48. Paul Best, "From Lab to Table: The Potential of Lab-Grown Meat and the Protectionist Push to Ban It," Cato Institute, *Free Society*, Winter 2024, https://www.cato.org/free
-society/winter-2024/lab-table-potential-lab-grown-meat-protectionist-push-ban-it; Jeffrey
A. Singer, "Governor DeSantis Tells Residents of the 'Free State of Florida' What Kind
of Meat He Will Allow Them to Eat," Cato Institute, May 3, 2024, https://www.cato
.org/blog/governor-desantis-tells-residents-free-state-florida-what-kind-meat-he-will-allow
-them-eat; Jacob Sullum, "Ron DeSantis Says Letting People Buy Cultivated Meat Is

Notes

Like Forcing Them To Eat Bugs," *Reason*, May 8, 2024, https://reason.com/2024/05/08/ron-desantis-says-letting-people-buy-cultivated-meat-is-like-forcing-them-to-eat-bugs/.

49. Nick Catoggio, "Meathead: Ron DeSantis' Terrible New New Right Policy," *Dispatch*, May 2, 2024, https://thedispatch.com/newsletter/boilingfrogs/meathead/.

50. Thomas Johnston, "Save the Beef?," Meatingplace, June 5, 2024, https://www.meatingplace.com/Industry/Blogs/Details/114837.

51. Jacob Ogles, "Meat Institute Argues Cellular Agriculture Ban Will Hurt Florida's Food Economy," *Florida Politics*, March 4, 2024, https://floridapolitics.com/archives/662944-meat-institute-argues-cellular-agriculture-ban-will-hurt-floridas-food-economy/.

52. *2017 Cattlemen's Stewardship Review* (National Cattlemen's Beef Association, 2017), https://www.beefitswhatsfordinner.com/Media/BIWFD/Docs/beef-csr-report-2017-final.pdf.

Chapter 10

1. Swanson, *Mitigating Risk and Capturing Opportunity*.

2. Matt Spence, "The National Security Case for Lab-Grown and Plant-Based Meat," *Slate*, December 29, 2021, https://slate.com/technology/2021/12/plant-cultivated-meat-national-security.html.

3. Terrence K. Kelly, Peter Chalk, James Bonomo, John V. Parachini, Brian A. Jackson, and Gary Cecchine, "The Office of Science and Technology Policy Blue Ribbon Panel on the Threat of Biological Terrorism Directed Against Livestock," (RAND Corporation, 2004), https://www.rand.org/pubs/conf_proceedings/CF193.html.

4. "Departing HHS Chief Cites Pandemic Flu, Food Tampering as Top Concerns," Center for Infectious Disease Research & Policy, https://www.cidrap.umn.edu/avian-influenza-bird-flu/departing-hhs-chief-cites-pandemic-flu-food-tampering-top-concerns.

5. Swanson, *Mitigating Risk and Capturing Opportunity*.

6. Zongyuan Zoe Liu, "China Increasingly Relies on Imported Food. That's a Problem," Council on Foreign Relations, January 25, 2023, https://www.cfr.org/article/china-increasingly-relies-imported-food-thats-problem.

7. Li Lei, "Nation Tops List of Agricultural Academic Papers, Patents," *China Daily*, last updated December 7, 2023, https://www.chinadaily.com.cn/a/202312/07/WS65712240a31090682a5f1e02.html; Yafei Li, Felix Herzog, Christian Levers, Franziska Mohr, Peter H. Verburg, Matthias Bürgi et al., "Agricultural Technology as a Driver of Sustainable Intensification: Insights From the Diffusion and Focus of Patents," *Agronomy for Sustainable Development* 44, article no. 14 (2024), https://doi.org/10.1007/s13593-024-00949-5.

8. Fred Gale and Fengxia Dong, *China's Meat Consumption: Growth Potential*, Report No. ERR-320 (US Department of Agriculture, Economic Research Service, 2023), https://search.nal.usda.gov/discovery/fulldisplay?context=L&vid=01NAL_INST:MAIN&docid=alma9916402832707426; Fred Gale and Fengxia Dong, "China Remains the World's Largest Meat Importer Despite Recent Declines," US Department of Agriculture, Economic Research Service, July 27, 2023, https://www.ers.usda.gov/data-products/charts-of-note/chart-detail?chartId=106996.

9. National Committee of the Chinese People's Political Consultative Conference, accessed June 15, 2025, http://en.cppcc.gov.cn/.

10. Genevieve Donnellon-May and Zhang Hongzhou, "Hungry China's Growing Interest

Notes

in 'Future Foods' and Alternative Protein," *Diplomat*, May 4, 2022, https://thediplomat .com/2022/05/hungry-chinas-growing-interest-in-future-foods-and-alternative-protein.

11. *China and Global Food Policy Report 2024: Building a Sustainable and Diversified Food Supply to Foster Agrifood Systems Transformation* (Academy of Global Food Economics and Policy, October 2024), https://agfep.cau.edu.cn/module/download/downfile.jsp?classid=0 &filename=e9e7cab381054fc0a1bdeb309d6548fb.pdf.

12. Dryer, "China Is on a Quest for the Holy Grail of Meat. Let's Hope It Succeeds," *New York Times*, November 28, 2024, https://www.nytimes.com/2024/11/28/opinion/china-food -science-protein.html.

13. "India at a Glance," Food and Agriculture Organization of the United Nations, accessed June 15, 2025, https://www.fao.org/india/fao-in-india/india-at-a-glance/en/; Banjot Kaur, "Globally, Highest Number of 'Undernourished' People Are In India: UN Report," *Wire*, July 2024, https://thewire.in/health/undernourished-highest-global-india-un-report.

14. Mayank Bhardwaj and Rajendra Jadhav, "Why Are Indian Farmers Protesting Again? Demands for Government Explained," *Reuters*, February 13, 2024, https://www.reuters .com/world/india/why-indian-farmers-are-marching-protest-capital-2024-02-13/; *Agricultural Policy Monitoring and Evaluation 2024: Innovation for Sustainable Productivity Growth* (OECD Publishing, 2025), https://doi.org/10.1787/74da57ed-en.

15. B. Pam Ismail, Lasika Senaratne-Lenagala, Alicia Stube, and Ann Brackenridge, "Protein Demand: Review of Plant and Animal Proteins Used in Alternative Protein Product Development and Production," *Animal Frontiers* 10, no. 4 (2020): 53–63, https://doi.org/10.1093 /af/vfaa040; Guoyao Wu, Jessica Fanzo, Dennis D. Miller, Prabhu Pingali, Mark Post, Jean L. Steiner et al., "Production and Supply of High-Quality Food Protein for Human Consumption: Sustainability, Challenges, and Innovations," *Annals of the New York Academy of Sciences* 1321, no. 1 (2014): 1–19, https://doi.org/10.1111/nyas.12500 ("Although some plant-based foods contain inhibitors that reduce protein digestibility, these inhibitors can be inactivated by adequate heat processing prior to consumption").

16. Rupam Jain, Rajendra Jadhav, and Manoj Kumar, "India's Farmer Protest Fuels Opposition Hopes of Denting Modi's Appeal," *Reuters*, February 29, 2024, https://www.reuters.com/world/ india/indias-farmer-protest-fuels-opposition-hopes-denting-modis-appeal-2024-02-29/.

17. "Pharmaceuticals," Invest India, accessed June 15, 2025, https://www.investindia.gov.in /sector/pharmaceuticals.

18. Ramya Ramamurthy, "Indian Government Grants Over $600,000 to Cell-Based Meat Research," Good Food Institute, April 26, 2019, https://gfi.org/blog/2019-04-26/. The only other government-funded cultivated meat research of which I'm aware were grants from NASA in 1995 and 1998 and the Dutch government in 2005. Lee et al., "Current Technology and Industrialization Status of Cell-Cultivated Meat."

19. Nikhil Inamdar and Aparna Alluri, "India Economy: Seven Years of Modi in Seven Charts," *BBC* June 21, 2021, https://www.bbc.com/news/world-asia-india-57437944.

20. "Why This Bipartisan Group of Senators Has Beef With the 'Big Meat' Industry," *Time*, April 28, 2022, https://time.com/6171326/meat-beef-industry-congress/.

21. Lydia Collas and Dustin Benton, *A New Land Dividend: The Opportunity of Alternative Proteins in Europe* (Green Alliance, May 9, 2024), https://green-alliance.org.uk/wp-content /uploads/2024/03/A_new_land_dividend.pdf (eight of the ten countries would be internally food self-sufficient. The UK and the Netherlands would still rely on land outside of their borders, but this could easily be covered by the surplus generated by the other 8 countries).

Notes

22. Mary Allen, "The Promise of Pulse Proteins," *Modern Farmer*, January 10, 2022, https://modernfarmer.com/2022/01/pulse-proteins/.
23. Evan A. Thaler, Isaac J. Larsen, and Qian Yu, "The Extent of Soil Loss Across the US Corn Belt," *Proceedings of the National Academy of Sciences* 118, no. 8 (2021), https://doi.org/10.1073/pnas.1922375118; Allen, "The Promise of Pulse Proteins."
24. "Carbon Dioxide Emissions From Domestic and International Aviation in the European Union (EU-27) From 1990 to 2022, by Type," Statistica, April 2024, https://www.statista.com/statistics/1306929/civil-aviation-co2-emissions-eu-by-type/.
25. See, e.g., Allen, "The Promise of Pulse Proteins."
26. *A Multi-Billion-Dollar Opportunity: Repurposing Agricultural Support to Transform Food Systems* (Food and Agriculture Organization of the United States, United Nations Development Programme, and United Nations Environment Programme, 2021), https://doi.org/10.4060/cb6562en; *Agricultural Policy Monitoring and Evaluation 2024* (OECD Publishing, 2025).
27. *The World Bank's Support for the Repurposing of Agrifood Public Policies and Programs* (World Bank, September 2024), https://thedocs.worldbank.org/en/doc/3da165e0bcb0ed7dddba9939afb21fda-0590012023/related/The-World-Bank-s-Support-for-Repurposing-of-Agrifood-Public-Policies-and-Programs-Sep-2024.pdf.
28. Sheryl Hendriks, Adrian de Groot Ruiz, Mario Herrero Acosta, Hans Baumers, Pietro Galgani, Daniel Mason-D'Croz et al., "The True Cost of Food: A Preliminary Assessment," in *Science and Innovations for Food Systems Transformation*, ed. Joachim von Braun, Kaosar Afsana, Louise O. Fresco, and Mohamed Hag Ali Hassan (Springer, 2023), https://doi.org/10.1007/978-3-031-15703-5_32, 581-601.
29. Robert F. Service, "U.S. Unveils Plans for Large Facilities to Capture Carbon Directly From Air," *ScienceInsider*, August 11, 2023, https://www.science.org/content/article/us-unveils-plans-for-large-facilities-to-capture-carbon-directly-from-air; Stephen M. Smith, Oliver Geden, Matthew J. Gidden, William F. Lamb, Gregory F. Nemet, Jan C. Minx et al., *The State of CDR: A Global, Independent Scientific Assessment of Carbon Dioxide Removal*, (Smith School of Enterprise and the Environment, University of Oxford, 2024), https://static1.squarespace.com/static/633458017a1ae214f3772c76/t/665ed1e2b9d34b2bf8e17c63/1717490167773/The-State-of-Carbon-Dioxide-Removal-2Edition.pdf.
30. Jan Dutkiewicz and Gabriel Rosenberg, "Labriculture Now," *Logic(s)*, no. 13, "Distribution," https://logicmag.io/distribution/labriculture-now/.
31. "Digital Report Launch," CSIS.
32. "JBS Is Entering the Cultivated Protein Market With the Acquisition of Bio Tech Foods and the Construction of a Plant in Europe," JBS, November 18, 2021, https://mediaroom.jbs.com.br/noticia/jbs-is-entering-the-cultivated-protein-market-with-the-acquisition-of-bio-tech-foods-and-the-construction-of-a-plant-in-europe ("JBS' entry into the cultivated protein market had the strategic support of The Good Food Institute, an entity that promotes plant and cell-based alternatives to animal products").
33. *Agricultural Policy Monitoring and Evaluation 2024* (OECD Publishing, 2025).
34. "Brasil Também Pode Ser Líder em Proteínas Alternativas, Apontam Especialistas No Setor" [Brazil Can Also Be a Leader in Alternative Proteins, Say Experts in the Sector], Brazil Ministry of Science, Technology and Innovation, June 28, 2023, https://www.gov.br/mcti/pt-br/acompanhe-o-mcti/noticias/2023/06/brasil-tambem-pode-ser-lider-em-proteinas-alternativas-apontam-especialistas-no-setor.
35. "R&D Spending Growth Slows in OECD, Surges in China; Government Support for

Notes

Energy and Defence R&D Rises Sharply," Organisation for Economic Co-operation and Development (OECD), March 31, 2025, https://www.oecd.org/en/data/insights/statistical -releases/2025/03/rd-spending-growth-slows-in-oecd-surges-in-china-government-support -for-energy-and-defence-rd-rises-sharply.html; Melissa Flagg and Paul Harris, *System Re-Engineering: A New Policy Framework for the American R&D System in a Changed World* (Center for Security and Emerging Technology, September 2020), https://cset.georgetown .edu/wp-content/uploads/CSET-System-Re-engineering.pdf; "Annual Articles Published in Scientific and Technical Journals," Our World in Data, last updated January 24, 2025, https:// ourworldindata.org/grapher/scientific-and-technical-journal-articles?tab=chart&country =GBR~CHE~IND~DEU~BRA~ISR~SGP~NLD~DNK~KOR~JPN; "Annual Patent Applications," Our World in Data, last updated January 24, 2025, https://ourworldindata .org/grapher/annual-patent-applications?tab=chart&country=DEU~JPN~IND~ISR ~BRA~SGP~USA~KOR; "World Intellectual Property Indicators Report: Record Number of Patent Applications Filed Worldwide in 2022," World Intellectual Property Organization, November 6, 2023, https://www.wipo.int/pressroom/en/articles/2023/article_0013.html.

36. "Trade Dependence of Agriculture and Food Products: Top Countries with Agriculture and Food Trade Surplus and Deficit," Econovis, June 12, 2024, https://www.econovis.net/post /trade-dependence-of-agriculture-and-food-products-top-countries-with-agriculture-and -food-trade-sur.

37. Baek Byung-yeul, "Korea's Grain Self-Sufficiency Lowest Among OECD Members," *Korea Times*, April 10, 2022, https://www.koreatimes.co.kr/www/tech/2022/04/419_327057.html; Yoona Jeon, *South Korea: Grain and Feed Annual*, Report KS2024-0014 (US Department of Agriculture, Foreign Agricultural Service, 2024), https://www.fas.usda.gov/data/south -korea-grain-and-feed-annual-8. Food self-sufficiency is a caloric analysis: What percent of calories is produced domestically? Grain self-sufficiency asks the same question about grain and is considered more important than food self-sufficiency, if a country is pushing up against famine or similar emergencies.

38. Singapore Food Statistics 2023 (Singapore Food Agency), January 2024, https://www.sfa .gov.sg/docs/default-source/publication/sg-food-statistics/singapore-food-statistics-2023.pdf.

39. "R&D Spending Growth Slows in OECD, Surges in China," OECD.

40. "CNA Explains: Where Does Singapore Get Its Food From?," *Channel News Asia*, May 27, 2022, https://www.channelnewsasia.com/singapore/cna-explains-where-does-singapore-get -its-food-2709161.

41. Judith Segaloff, "'Israel Has to Be Much More Food Self-Sufficient,' Expert Warns," *Jewish News Syndicate*, January 29, 2024, https://www.jns.org/israel-has-to-be-much-more -food-self-sufficient-expert-warns; Oren Shaked, *An Overview of Israeli Beef Market*, Report KS2024-0024 (US Department of Agriculture, Foreign Agricultural Service, 2024), https:// www.fas.usda.gov/data/israel-overview-israeli-beef-market.

42. Allyson Chai, "No More Chicken: How Poultry Shortages Give Us a Taste of Singapore's Complex Stance in International Politics," Roosevelt Group, October 31, 2023, https://www .roosevelt-group.org/quick-takes/singapore-chicken-shortage.

43. "Fatwa on Cultivated Meat," Majlis Ugama Islam Singapura, February 3, 2024, https://www .muis.gov.sg/resources/media-releases/3-feb-24-fatwa-on-cultivated-meat.

44. Anay Mridul, "Korean Muslim Group Becomes Latest Religious Organization to Issue a Fatwa Approving Cultivated Meat as Halal," Genetic Literacy Project, March 26, 2025, https://

Notes

geneticliteracyproject.org/2025/03/26/korean-muslim-group-becomes-latest-religious
-organization-to-issue-a-fatwa-approving-cultivated-meat-as-halal/.

45. Melissa Sue Sorrells, "Islamic Authority Rules on Cultivated Meat," Alt-Meat, June 3, 2025, https://www.alt-meat.net/islamic-authority-rules-cultivated-meat.
46. Ryan Huling, Doris Lee, Wasamon Nutakul, and Samuel Goh, "In Asia, Alternative Proteins Are the New Clean Energy," *Nature* 633, no. 8031 (2024): 767-769, https://doi.org/10.1038/d41586-024-03077-y.
47. Huling et al., "In Asia, Alternative Proteins Are the New Clean Energy"; Yeshi Liang and Doris Lee, "Recent Progress of Cultivated Meat in Asia," *Food Materials Research* 2, no. 1 (2022): 1–8, https://doi.org/10.48130/fmr-2022-0012; Megumi Avigail Yoshitomi, "Japan Association for Cellular Agriculture: 'Working to Achieve the Sale of Cultivated Meat in Japan from 2026,'" Cultivated X, February 8, 2024, https://cultivated-x.com/interviews/japan-association-for-cellular-agriculture-working-achieve-sale-cultivated-meat-japan-2026/.
48. Steve Brawner, "Engineers Questioned Moon Shot, but Solved Problems Through Clarity, Ownership," *Talk Business & Politics*, October 20, 2019, https://talkbusiness.net/2019/10/engineers-questioned-moon-shot-but-solved-problems-through-clarity-ownership/.
49. Nemet, *How Solar Energy Became Cheap*, 31.
50. Milad Haghani, Frances Sprei, Khashayar Kazemzadeh, Zahra Shahhoseini, Jamshid Aghaei, "Trends in Electric Vehicles Research," *Transportation Research Part D: Transport and Environment* 123 (2023): 103881, https://doi.org/10.1016/j.trd.2023.103881.

Chapter 11

1. Good Meat and Vow in Singapore; Good Meat, Upside, Mission Barns, and Wildtype in the United States; Aleph Farms in Israel; and Vow in Australia and New Zealand. The first was Good Meat, in Singapore, in December 2020. Then Good Meat and Upside Foods received regulatory approval in the United States in June 2023. Upside sold its meat at Bar Crenn in San Francisco, and Good Meat sold its meat at another of José Andrés' D.C. restaurants, China Chilcano. The Israeli company Aleph Farms also received approval to sell cultivated steak in Israel, but as I'm writing in late spring 2025, it has not yet done so. "Human Food Made with Cultured Animal Cells Inventory," US Food & Drug Administration, last updated June 2025, https://www.hfpappexternal.fda.gov/scripts/fdcc/index.cfm?set=AnimalCellCultureFoods. Jennifer L., "Climate Tech VC Investments Drop for the 3rd Year in a Row, AI to the Rescue," Carbon Credits, January 22, 2025, https://carboncredits.com/climate-tech-vc-investments-drop-for-the-3rd-year-in-a-row-ai-to-the-rescue/; Krystal Hu, "US Startup Funding Drops 30% in 2023 Despite AI Frenzy," *Reuters*, January 11, 2024, https://www.reuters.com/technology/us-startup-funding-continues-drop-despite-ai-frenzy-2024-01-11/.
2. Jennifer L., "Climate Tech VC Investments Drop for the 3rd Year in a Row, AI to the Rescue," Carbon Credits, January 22, 2025, https://carboncredits.com/climate-tech-vc-investments-drop-for-the-3rd-year-in-a-row-ai-to-the-rescue/; Krystal Hu, "US Startup Funding Drops 30% in 2023 Despite AI Frenzy," *Reuters*, January 11, 2024, https://www.reuters.com/technology/us-startup-funding-continues-drop-despite-ai-frenzy-2024-01-11/.
3. Battle et al. *2024 State of the Industry*, 16.
4. "UPSIDE Foods Raises a $400M Series C Round to Commercialize Cultivated Meat at Scale," PR Newswire, April 21, 2022, https://www.prnewswire.com/news-releases/upside-foods

Notes

-raises-a-400m-series-c-round-to-commercialize-cultivated-meat-at-scale-301529998.html ("UPSIDE Foods has raised a total of $608 million").

5. Olivier J. Wouters, Martin McKee, and Jeroen Luyten, "Estimated Research and Development Investment Needed to Bring a New Medicine to Market, 2009–2018," *Journal of the American Medical Association*, 323, no. 9 (2020): 844–853, https://doi.org/10.1001/jama .2020.1166.

6. River Davis, "Toyota to Scale Up Battery Output," *Wall Street Journal*, September 1, 2022, B3; Rebecca Elliott, "Redwood Materials Plans $3.5 Billion Battery-Materials Plant in Nevada," *Wall Street Journal*, last updated July 25, 2022, https://www.wsj.com/business/autos/ redwood-materials-plans-3-5-billion-battery-materials-plant-in-nevada-11658750400?mod =Searchresults_pos5; Tim Higgins, "Big US Battery Plant Advances," *Wall Street Journal*, October 6, 2022, B3; William Boston, "U.K. Clinches $5 Billion EV-Battery Plant as Race for Green Investments Heats Up," *Wall Street Journal*, July 19, 2023, https://www.wsj.com /business/u-k-cinches-5-billion-ev-battery-plant-as-race-for-green-investments-heats-up -e20b142f (1. Battery factory discussion: $4.4 billion factory in the US (Honda), plus two more similarly sized in Kansas and Oklahoma (Tesla). 2. Factory in NW Nevada: $3.5 billion factory. 3. $4 billion factory in Kansas (Panasonic). 4. Tata (Indian company): $5.2 billion factory in the UK).

7. "Digital Report Launch," CSIS.

8. Sam Bourgi, "The End of the Electric Vehicle Startup. 8 of the 10 Top EV Makers Are Legacy Automakers," *Investors Observer*, April 23, 2025, https://investorsobserver.com/news /stock-update/the-end-of-the-electric-vehicle-startup-8-of-the-10-top-ev-makers-are-legacy -automakers/; Cassandra Cassidy, "Nikola, a Once-Hyped EV-Maker, Has Gone Bankrupt," *Morning Brew*, February 19, 2025, https://www.morningbrew.com/stories/2025/02/20 /nikola-a-once-hyped-ev-maker-has-gone-bankrupt; Robert Ferris, "Here's Why So Many Electric Vehicle Startups Fail," CNBC, April 18, 2024, https://www.cnbc.com/2024/04/18 /why-so-many-electric-vehicle-startups-fail.html; Paul A. Eisenstein, "Electric Car Start-Ups — Once Seen as Threat — Now Struggle to Survive," CNBC, October 19, 2019, https:// www.cnbc.com/2019/10/19/electric-car-start-ups-once-seen-as-threat-now-struggle-to -survive.html.

9. Elaine Watson, "Cultivated Meat at a Crossroads: Highlights From the Tufts Cell Ag Innovation Day," *AgFunder News*, January 10, 2025, https://agfundernews.com/cultivated-meat -at-a-crossroads-highlights-from-the-tufts-cell-ag-innovation-day (New Age Meats had a lead investor who wanted them to change their focus, which they refused to do; SciFi Foods had a lead investor that had a change in priorities; in both cases, the companies had accepted investments from too few different investors to recover).

10. Extrapolating from the IIASA *Nature Communications* analysis, which found that at 90% plant-based meat, climate benefits equal 11.9 Gt. Marta Kozicka, Petr Havlík, Hugo Valin, Eva Wollenberg, Andre Deppermann, David Leclère et al., "Feeding Climate and Biodiversity Goals With Novel Plant-Based Meat and Milk Alternatives," *Nature Communications* 14, article no. 5316 (2023), https://doi.org/10.1038/s41467-023-40899-2.

11. "Yearly Number of Animals Slaughtered for Meat, United States, 1961 to 2023," Our World in Data, last updated March 17, 2025, https://ourworldindata.org/grapher /animals-slaughtered-for-meat?country=~USA.

12. Derya Yildiz, "North America Animal Feed Market," *Feed & Additive*, September 12, 2023, https://www.feedandadditive.com/north-america-animal-feed-market; "Feed Facts,"

Notes

American Feed Industry Organization, accessed June 15, 2025, https://www.afia.org/feedfacts/feed-industry-stats/.

13. Amy R. Sapkota, Lisa Y. Lefferts, Shawn McKenzie, and Polly Walker, "What Do We Feed to Food-Production Animals? A Review of Animal Feed Ingredients and Their Potential Impacts on Human Health," *Environmental Health Perspectives* 115, no. 5 (2007): 663-670, https://doi.org/10.1289/ehp.9760; "Ingredients & Additives," US Food & Drug Administration, last updated January 31, 2024, https://www.fda.gov/animal-veterinary/animal-food-feeds/ingredients-additives.

14. "Total Meat Production," Our World in Data, accessed June 12, 2025, https://ourworldindata.org/grapher/meat-production-tonnes?tab=chart; *Livestock Slaughter 2023 Summary* (US Department of Agriculture, April 2024), https://downloads.usda.library.cornell.edu/usda-esmis/files/r207tp32d/wh248d422/p5549g65c/lsan0424.pdf; *Poultry Slaughter* (US Department of Agriculture, May 2024), https://downloads.usda.library.cornell.edu/usda-esmis/files/3197xm04j/ht24z779c/jd474k352/psla0524.pdf.

15. "Impacts of climate change on global agriculture accounting for adaptation," *Nature*, June 18, 2025, https://www.nature.com/articles/s41586-025-09085-w.

16. "A New Frontier for Global Energy Security: Critical Minerals," International Energy Agency, accessed June 15, 2025, https://www.iea.org/topics/critical-minerals.

17. "The Role of Critical Minerals in Clean Energy Transitions: Executive Summary," in *World Energy Outlook* (International Energy Agency, May 2021), https://www.iea.org/reports/the-role-of-critical-minerals-in-clean-energy-transitions/executive-summary; "World Steel in Figures 2024," World Steel Association, accessed June 15, 2025, https://worldsteel.org/data/world-steel-in-figures/world-steel-in-figures-2024/.

18. Laura Pasitka, Guy Wissotsky, Muneef Ayyash, Nir Yarza, Gal Rosoff, Revital Kaminker et al., "Empirical Economic Analysis Shows Cost-Effective Continuous Manufacturing of Cultivated Chicken Using Animal-Free Medium," *Nature Food* 5 (2004): 693-702, https://doi.org/10.1038/s43016-024-01022-w; Elaine Watson, "The Death of Cultivated Meat Has Been Greatly Exaggerated, Says Report as Vow Predicts It Will Soon Be 'Unit Margin Positive'," *AgFunder News*, April 10, 2025, https://agfundernews.com/humbird-was-spectacularly-wrong-on-cultivated-meat-economics-says-report-as-vow-predicts-it-will-soon-be-unit-margin-positive; Elaine Watson, "Biosphere Emerges from Stealth, Raises $8.8m for UV-Sterilized Bioreactors It Claims Can Slash Bioproduction Costs," *AgFunder News*, January 9, 2025, https://agfundernews.com/biosphere-emerges-from-stealth-raises-8-8m-for-uv-sterilized-bioreactors-it-claims-can-slash-bioproduction-costs.

19. "9 Ways AI Is Advancing Science," Google Blog, November 18, 2024, https://blog.google/technology/ai/google-ai-big-scientific-breakthroughs-2024/.

20. Anil Ananthaswamy, "AI Designs Quantum Physics Experiments beyond What Any Human Has Conceived," *Scientific American*, July 2, 2021, https://www.scientificamerican.com/article/ai-designs-quantum-physics-experiments-beyond-what-any-human-has-conceived/.

21. Sara Frueh, "How AI Is Shaping Scientific Discovery," National Academies of Science, Engineering, and Medicine, November 6, 2023, https://www.nationalacademies.org/news/2023/11/how-ai-is-shaping-scientific-discovery.

22. Paul Nyhan, "2 AI Breakthroughs Unlock New Potential for Health and Science," Microsoft, January 17, 2025, https://news.microsoft.com/source/features/ai/2-ai-breakthroughs-unlock-new-potential-for-health-and-science/.

23. "Bezos Earth Fund Announces 24 Phase I Grants Under AI Grand Challenge for

Notes

Climate and Nature," Bezos Earth Fund, May 21, 2025, https://www.bezosearthfund.org/news-and-insights/phase-i-grants-ai-grand-challenge-climate-nature.

24. Watson, "Death of Cultivated Meat Has Been Greatly Exaggerated" (email from George Peppou updates from 20,000 to 22,000 liters).

25. Winston Churchill, "Fifty Years Hence," *Strand Magazine*, December 1931.

26. Ezra Klein, "Let's Launch a Moonshot for Meatless Meat," *New York Times*, April 24, 2021, https://www.nytimes.com/2021/04/24/opinion/climate-change-meatless-meat.html.

27. Battle et al. *2024 State of the Industry.*

28. *Global Food System* (Credit Suisse, June 2021), 60, figure 28.

29. Battle et al. *2024 State of the Industry.*

30. "State Motor Vehicle Registrations, by Years, 1900–1995," US Department of Transportation, https://www.fhwa.dot.gov/ohim/summary95/mv200.pdf.

31. "Global EV Data Explorer," IEA.

32. "Batteries and Secure Energy Transitions: Executive Summary," in *World Energy Outlook 2023* (International Energy Agency, April 2024), https://www.iea.org/reports/batteries-and-secure-energy-transitions/executive-summary.

Conclusion

1. Peter Coy, "Why Don't We Just Ban Fossil Fuels?," *New York Times*, February 16, 2024, https://www.nytimes.com/2024/02/16/opinion/fossil-fuel-ban-climate.html.

Index

Index

Index

Index

Index

Index

Index

Index

Rosenberg, Gabriel, 210, 242
Rowat, Amy, 247
Rubio, Natalie, 113, 247–248
Russia, xii, 17, 59
Rwanda, 19–20

S

safety issues, 20, 98, 104, 116, 119–124, 193
saturated fat, 133–134, 136
Schattenmann, Florian, 191
Schmidt, Eric, 138
Schulze, Eric, 113
science, government, and industry partnerships, 171–185, 249
  in agriculture, 180–182
  on cultivated meat, 114–115, 185
  on electric vehicles, 176–177
  for military and space applications, 185
  in pharmaceutical industry, 179–180
  on photovoltaics, 175–176, 178
  on plant-based meat, 146–149
  to scale up alt meat production, 238, 242
  in technology industry, 182
  as three-legged stool of innovation, 6–7, 171–175
scientists, 140–141, 247–249
SciFi Foods, 226
Sculley, John, 155, 167
Scully, Matthew, 91
S-curve of adoption, 49, 161, 169, 241
seafood, 47–48, 122–124, 183, 215, 234–235
self-sufficiency, 199–202, 206, 213

Sen, Amarytya, 16
Shanmugaratnam, Tharman, 214
Shapiro, Paul, 92, 105, 151
Shortridge, Kennedy, 66
Singapore, 147–148, 186, 212–214, 221, 233, 234
Singer, Peter, 91, 92
smallholder and family farms, 16, 19–21, 205–206, 208, 210
smartphones, 2, 157, 169
Smith, Jimmy, 75
solar energy, 8, 29, 160–163, 175–176, 178, 187, 217
South Korea, 212–213, 215–216, 233
soy, 22–23, 45–47, 141, 143, 145–147, 149–151, 202, 205–206, 208
SpaceX, 110, 185, 234, 241
Specht, Liz, 54, 107–111, 112n*, 233
Spence, Matt, 187, 195–197, 216–217, 224
startups, 105–106, 141–143, 175, 222–226, 251–252
steel, 227, 229
Stein, Dan, 255–256
Stone, Joe, 106n*
sugar, 133, 135–137
superbugs, 52, 54–57, 60
supply chain issues, 75, 197–198, 235–236
Swanson, Zane, 197
Swartz, Elliot, 114, 240–241
swine flu, 69–70
system one and two thinking, 95–96

T

Tabuchi, Hiroko, 123
taste parity
  for alternative meat, 49, 93, 99, 101, 191, 221
  for blended meats, 149

  for cultivated meat, 29, 116–119, 235
  for plant-based meat, 29, 126–127, 130–132, 138, 139, 144–146, 223, 239
Tayag, Yasmin, 92, 96
Teach for America, 8, 69, 82, 259
technological innovations, 155–170
  adoption of, 3
  airplanes, 166–167
  alternative meats, 138–140
  artificial ice, 165–166
  automobiles, 157–159
  cultivated meat, 174–175
  electric vehicles, 160–162
  and failed alt meat companies, 224
  food security strategies of nations that excel at, 212–216
  home computers, 167
  Human Genome Project, 163–165
  and limits of human imagination, 156–157
  Moore's, Wright's, and Amara's Laws for, 162–163
  online shopping, 168
  to reduce hunger, 20–21
  to scale up alt meat production, 8–9
  smartphones, 2
  solar energy, 160–162
  three-legged stool of, 6–7, 171–175
technology industry, 182, 238–239
Tetrick, Josh, 119
texturized vegetable protein (TVP), 19–20
think tanks, 252–253

Index

About the Author

Bruce Friedrich is the founder and president of the Good Food Institute (GFI), a global science think tank with more than 230 full-time team members, the plurality scientists. *New York Times* columnist Ezra Klein writes that GFI is "second-to-none in the influence of its public policy efforts, its centrality to the ecosystem of companies and researchers, and its international footprint. It has also been effective at convincing traditional meat companies to explore alternative proteins, which could lead both to important products and turn political enemies into allies."

Bruce has written for *The Wall Street Journal*, *Foreign Policy*, *Wired*, *Nature*, and more. His TED Talk has been viewed more than 2.4 million times and translated into 30 languages. Bruce graduated magna cum laude from the Georgetown University Law Center and holds degrees from Johns Hopkins University, London School of Economics, and Grinnell College.